Offshore Operations and Engineering

Offshore Operations and Engineering

Shashi Shekhar Prasad Singh,
Jatin R. Agarwal, and Nag Mani

CRC Press is an imprint of the
Taylor & Francis Group, an **informa** business

CRC Press
Taylor & Francis Group
6000 Broken Sound Parkway NW, Suite 300
Boca Raton, FL 33487-2742

© 2020 by Taylor & Francis Group, LLC
CRC Press is an imprint of Taylor & Francis Group, an Informa business

No claim to original U.S. Government works

Printed on acid-free paper

International Standard Book Number-13: 978-0-3673-7434-1 (Hardback)

This book contains information obtained from authentic and highly regarded sources. Reasonable efforts have been made to publish reliable data and information, but the author and publisher cannot assume responsibility for the validity of all materials or the consequences of their use. The authors and publishers have attempted to trace the copyright holders of all material reproduced in this publication and apologize to copyright holders if permission to publish in this form has not been obtained. If any copyright material has not been acknowledged, please write and let us know so we may rectify in any future reprint.

Except as permitted under U.S. Copyright Law, no part of this book may be reprinted, reproduced, transmitted, or utilized in any form by any electronic, mechanical, or other means, now known or hereafter invented, including photocopying, microfilming, and recording, or in any information storage or retrieval system, without written permission from the publishers.

For permission to photocopy or use material electronically from this work, please access www.copyright.com (http://www.copyright.com/) or contact the Copyright Clearance Center, Inc. (CCC), 222 Rosewood Drive, Danvers, MA 01923, 978-750-8400. CCC is a not-for-profit organization that provides licenses and registration for a variety of users. For organizations that have been granted a photocopy license by the CCC, a separate system of payment has been arranged.

Trademark Notice: Product or corporate names may be trademarks or registered trademarks, and are used only for identification and explanation without intent to infringe.

Visit the Taylor & Francis Web site at
http://www.taylorandfrancis.com

and the CRC Press Web site at
http://www.crcpress.com

Contents

Foreword xvii

Preface...xix

Acknowledgments..xxi

Authors..xxiii

Chapter 1   Introduction to Offshore Operation...1

    1.1   Ocean Baseline..1

    1.2   Ocean Environment...2

        1.2.1   Metocean Condition and Study......................................2

        1.2.2   Meteorology...3

        1.2.3   Physical Oceanography..3

        1.2.4   Metocean Data...4

    1.3   Offshore Oil and Gas Production..4

        1.3.1   Comparison of Onshore versus
Offshore Oil Production...5

        1.3.2   Comparision of Offshore Oil Production on the
Basis of Water Depth...6

        1.3.3   Rig Count and Utilization Rate......................................7

    1.4   Offshore Operations..7

        1.4.1   Notable Offshore Fields across the Globe..................10

        1.4.2   Major Offshore Oil Producing Countries...................10

        1.4.3   Offshore Facilities (Installations/Platforms/Rigs).....11

        1.4.4   Types of Offshore Installations/Platforms.................11

        1.4.5   Type of Offshore Rigs...11

        1.4.6   Challenges with Offshore Facilities............................11

        1.4.7   Ecological and Environmental Risks...........................12

    1.5   How Are Offshore Operations Different from Onshore.........13

        1.5.1   Safety Requirements..13

    1.6   Technology Wave...13

    References..14

Chapter 2   Offshore Structure and Design...15

    2.1   Structural Designing..16

        2.1.1   Corrosion Protection..17

        2.1.2   Cathodic Protection...17

               2.1.2.1   Impressed Current Cathodic
Protection Systems....................................18

               2.1.2.2   Sacrificial Anodes (Galvanic Action).........18

        2.1.3   Loads...19

               2.1.3.1   Constant Loads...19

v

		2.1.3.2	Variable Loads 19
		2.1.3.3	Environment-Dependent Loads (Normally Calculated on Historically 100-Year Return Period) 20
		2.1.3.4	Additional Loads during Installation and Construction 22
		2.1.3.5	Unforeseen Load 22

2.2 Fixed Platforms

2.2 Fixed Platforms 22

2.2.1 Concrete Gravity Structure 22

    2.2.1.1 Design 25

    2.2.1.2 Construction and Installation 26

2.2.2 Compliant Tower and Guyed Tower 27

    2.2.2.1 Design 27

    2.2.2.2 Construction and Installation 28

2.2.3 Jacketed Platform 30

2.2.4 Jackup Platform/Rig 33

    2.2.4.1 Design 33

    2.2.4.2 Installation 34

2.3 Floating Platform 35

2.3.1 Mooring and Anchoring 35

    2.3.1.1 Classification of Mooring Systems 38

    2.3.1.2 Catenaries Mooring System 38

    2.3.1.3 Single Point Mooring System 39

References 40

Chapter 3 Offshore Drilling and Completion 43

3.1 Offshore Drilling 43

3.1.1 Introduction 43

3.1.2 Well Planning 44

    3.1.2.1 Geology 45

    3.1.2.2 Completion Design 45

    3.1.2.3 Trajectory Design 45

    3.1.2.4 Wellbore Stability 46

    3.1.2.5 Drilling Fluid Design 46

    3.1.2.6 Casing Design 46

    3.1.2.7 Cement Job Designs 46

    3.1.2.8 Wellhead/Blowout Prevention (BOP) Design 46

    3.1.2.9 BHA and Drill String 46

    3.1.2.10 Bit Design 46

    3.1.2.11 Hole Cleaning and Hydraulics 46

    3.1.2.12 Rig Equipment 46

3.1.3 Rig Selection Criteria 47

    3.1.3.1 HSE Compatibility 47

    3.1.3.2 Technical Capability 47

Contents vii

		3.1.3.3	Full-Cycle Efficiency	47
	3.1.4	Wellbore Stability		47
		3.1.4.1	Stress Distribution around the Wellbore	48
		3.1.4.2	Establishing a Minimum Safe Mud Weight	49
		3.1.4.3	Validating the Geomechanical Model	50
		3.1.4.4	Sand Production	51
	3.1.5	Casing Design		51
		3.1.5.1	Casing Program	51
		3.1.5.2	Loads Encountered While Designing Casing	52
	3.1.6	Trajectory Design		57
		3.1.6.1	The Target	58
		3.1.6.2	Kick-Off Point and Build-Up Rate	58
		3.1.6.3	Tangent Section	59
		3.1.6.4	Drop-Off Section	59
		3.1.6.5	Trajectory Measurements	59
	3.1.7	Directional Drilling		64
		3.1.7.1	Evolution of Directional Drilling	64
		3.1.7.2	Types of Directional Wells	64
		3.1.7.3	Conventional Drilling versus Directional Drilling	65
		3.1.7.4	Directional Drilling Tools	65
		3.1.7.5	Measurement While Drilling (MWD) Tool	65
		3.1.7.6	Importance and Uses of Directional Drilling	66
		3.1.7.7	Application of Directional Drilling	67
	3.1.8	Dual Gradient Drilling		68
		3.1.8.1	Introduction	68
		3.1.8.2	Dual Gradient Drilling – Overview	68
		3.1.8.3	Single versus Dual Gradient Drilling	68
		3.1.8.4	Dual Gradient System	69
		3.1.8.5	Types of Dual Gradient Drilling	70
		3.1.8.6	Limitations of DGD	71
3.2	Offshore Well Completion			72
	3.2.1	Introduction		72
	3.2.2	Well Completion Concepts in Offshore		73
		3.2.2.1	Classification of Completions	73
	3.2.3	Horizontal Well Completions		80
		3.2.3.1	Open Hole Completion	80
		3.2.3.2	Slotted Liner Completion	80
		3.2.3.3	Slotted Liner Completion for Zonal Isolation	81
		3.2.3.4	Cemented and Perforated Completion	82
	3.2.4	Intelligent Well Systems		83

| | 3.2.5 | Multilateral Completions | 83 |

3.2.5 Multilateral Completions...83
    3.2.5.1 Multilateral Well Classification.................84
    3.2.5.2 Level 1 Multilateral Well.........................84
    3.2.5.3 Level 2 Multilateral Well.........................85
    3.2.5.4 Level 3 Well Completions........................86
    3.2.5.5 Level 4 Well Completions........................86
    3.2.5.6 Level 5 Well Completions........................86
    3.2.5.7 Level 6 Well Completions........................86
3.2.6 Subsea Completion...86
    3.2.6.1 Types of Subsea Completions...................87
3.2.7 Completion Equipment..89
    3.2.7.1 Christmas Tree (Xmas Tree).....................89
    3.2.7.2 Production Tubing..................................89
    3.2.7.3 Packers...90
    3.2.7.4 Blast Joint..92
    3.2.7.5 Flow Coupling.....................................92
    3.2.7.6 Seating Nipple.....................................92
    3.2.7.7 Landing Nipple....................................93
    3.2.7.8 Expansion Joint...................................93
    3.2.7.9 Safety Joints.......................................93
    3.2.7.10 Safety Valves.....................................93
    3.2.7.11 Circulating Valves..............................95
References..96

Chapter 4 Offshore Oil and Gas Production and Transportation.......................99

4.1 Offshore Production Operation..99
  4.1.1 Major Elements of Offshore Production System......100
    4.1.1.1 Wells (Subsea/Platform Wells).................100
    4.1.1.2 Platform Wells/Dry Trees.......................100
    4.1.1.3 Subsea Wells/Wet Trees.........................100
    4.1.1.4 Offshore Pipelines................................101
    4.1.1.5 Processing Platforms.............................101
    4.1.1.6 Export Pipelines/Tankers for Evacuation of Oil and Gas......................102
  4.1.2 Maintenance and Supply...103
  4.1.3 Essential Personnel/Workforce...............................103
  4.1.4 Risks...105
    4.1.4.1 Basic Protection Concepts.......................105
  4.1.5 Prevention..105
  4.1.6 Shut In..107
    4.1.6.1 Fire and Gas Leakage Protection System..................................107
    4.1.6.2 Technology Development: HIPS.............107
    4.1.6.3 Surface Facility Protection......................107
    4.1.6.4 Well Control and Protection....................109

Contents ix

| | 4.1.7 | SCADA – An Essential Part of Digital Oil and Gas Field | 109 |

       4.1.7 SCADA – An Essential Part of Digital
       Oil and Gas Field 109
          4.1.7.1 Process Levels in SCADA 109
          4.1.7.2 Instrumentation, Remote Sensing, and
                  Telemetry of Real-Time Processes 111
       4.1.8 Automated Gas Lift Optimization in Offshore 114
   4.2 Processing in Offshore 116
       4.2.1 Oil Treatment 116
          4.2.1.1 Important Notes 118
          4.2.1.2 Loading of Tankers 118
       4.2.2 Gas Treatment 118
          4.2.2.1 Gas Dehydration Follows the Steps Below ... 119
       4.2.3 Produced Water Treatment 120
   4.3 Sea Water Injection 121
   4.4 Offshore Storage 122
   4.5 Transportation of Oil and Gas 123
       4.5.1 Oil Tankers 123
       4.5.2 Pipelines 124
       4.5.3 Floating Production, Storage, and
          Offloading (FPSO) 124
          4.5.3.1 Single Buoy Mooring 125
  References 126

Chapter 5 Utilities and Support System 127

   5.1 Living Accommodation 127
       5.1.1 Medical 128
       5.1.2 Smoking and Alcohol 129
       5.1.3 Entertainment and Recreation 129
   5.2 Power Generation 129
       5.2.1 Fuel Gas System 129
       5.2.2 Utility/Diesel Generators 130
       5.2.3 Gas Turbine Generator 130
   5.3 Instrument and Utility Air System 132
       5.3.1 Air Compressors 132
       5.3.2 Instrument Air and Utility Air Systems 132
   5.4 Hot Oil System 133
       5.4.1 Crude Oil Heater 133
       5.4.2 Chemical Tank 133
       5.4.3 Glycol Reboiler 133
       5.4.4 Skimmer Vessel 133
   5.5 Potable Water System 133
   5.6 Water Cooling System 134
       5.6.1 Freshwater Cooling System 134
   5.7 Utility Water System 135
   5.8 Drain Header and Sump Caisson 135

x Contents

	5.9	Heating, Ventilation, and Air Conditioning Equipment	136
	5.10	Communication System	136
		5.10.1 Satellite	136
		5.10.2 Microwave Telecommunication	137
		5.10.3 Optical Fibers	137
		5.10.4 Cellular Services	137
	5.11	Diesel System	137
	5.12	Sewage Treatment System	138
	5.13	Material Handling	138
	5.14	Offshore Logistics	139
		5.14.1 Air Logistics	139
		5.14.2 Sea Logistics	139
	References		142

Chapter 6	Deep Sea Development		145
	6.1	Factors Driving Deep Sea Development	145
	6.2	Deep Sea Development Options	145
		6.2.1 Recoverable Reserves	145
		6.2.2 Water Depth	147
		6.2.3 Challenges in Subsea due to Water Depth	147
		6.2.4 Production Rate	147
		6.2.5 Reservoir Structure	148
		6.2.6 Reservoir Production Characteristics	148
		6.2.7 Environmental and Geological Conditions	148
		6.2.8 Existing Infrastructure	149
	6.3	Subsea Field Development	149
		6.3.1 Subsea Well Completion	151
		6.3.2 Subsea Christmas Tree	151
		6.3.2.1 Dry Tree Systems	151
		6.3.2.2 Wet Tree Systems	153
		6.3.3 Subsea Tieback Development	153
		6.3.3.1 Challenges	153
		6.3.3.2 Stand-Alone Development	155
		6.3.3.3 Well Groupings	155
		6.3.3.4 Satellite Well System	155
		6.3.3.5 Template Well System	156
		6.3.3.6 Clustered Well System	157
		6.3.3.7 Production Well Templates	157
		6.3.3.8 Daisy Chain	157
		6.3.3.9 Subsea Monitoring, Control, and Communication System	158
		6.3.4 Main Topside Elements	159
		6.3.4.1 MCS	159
		6.3.4.2 Electrical Power Unit (EPU)	160
		6.3.4.3 HPU	160

Contents | xi

6.3.5   Topside Umbilical Termination Assembly (TUTA).... 160
6.4   Subsea Elements 162
    6.4.1   SDS Components 163
        6.4.1.1   Umbilical 163
        6.4.1.2   Subsea Umbilical Termination
                     Assembly (SUTA) 163
        6.4.1.3   Umbilical Termination Head (UTH) 165
        6.4.1.4   Subsea Distribution Assembly 165
        6.4.1.5   Hydraulic Distribution Manifold/
                     Module (HDM) 165
        6.4.1.6   Electrical Distribution Manifold/
                     Module 167
        6.4.1.7   Multiple Quick Connects 167
        6.4.1.8   Hydraulic Flying Leads (HFL) 168
        6.4.1.9   Hydraulic Couplers 168
        6.4.1.10 Electrical Flying Leads 169
        6.4.1.11 Logic Caps 169
        6.4.1.12 Subsea Accumulator Module 169
        6.4.1.13 Subsea Control Module 169
        6.4.1.14 Transducer/Sensor 171
        6.4.1.15 Subsea Production Control System 171
        6.4.1.16 Types of Control Systems 173
6.5   Subsea Power Supply 179
6.6   Flow Assurance 179
    6.6.1   Shallow versus Deep Flow Assurance Scenario 180
    6.6.2   Flow Assurance Challenges 180
    6.6.3   Troublemakers 180
        6.6.3.1   Gas Hydrates 180
        6.6.3.2   Paraffin/Wax 181
        6.6.3.3   Asphaltene 181
        6.6.3.4   Scales 181
        6.6.3.5   Erosion 181
        6.6.3.6   Corrosion 181
        6.6.3.7   Slugging 182
        6.6.3.8   Severe Slugging 182
    6.6.4   Typical Flow Assurance Processes 182
    6.6.5   Fluid Characterization and Flow Property
        Assessments 182
    6.6.6   Steady-State Hydraulic and Thermal
        Performance Analyses 183
    6.6.7   System Design and Operability 184
        6.6.7.1   System Start-Up 184
        6.6.7.2   System Shutdown 184
    6.6.8   Transient Flow Hydraulic and Thermal
        Performance Analyses 185
    6.6.9   Hydrate Prevention Methods 186

		6.6.9.1	Thermodynamic Inhibitors	186
		6.6.9.2	Low Dosage Inhibitors (LDIs)	187
		6.6.9.3	Kinetic Inhibitors (KIs)	187
		6.6.9.4	Anti-Agglomerates (AAs)	187
		6.6.9.5	Low-Pressure Operations	187
		6.6.9.6	Water Removal	187
		6.6.9.7	Thermal Insulation and Heating	187
	6.6.10	Hydrate Remediation		188
	6.6.11	Selection of Hydrate Control Method		189
		6.6.11.1	Gas System	189
		6.6.11.2	Oil System	189
	6.6.12	Wax Control Guidelines		189
	6.6.13	Wax Management Strategy		189
		6.6.13.1	Thermal Control	189
		6.6.13.2	Chemical Inhibition	189
		6.6.13.3	Thermal and Chemical Wax Dissolution	190
		6.6.13.4	Physical Removal	190
	6.6.14	Asphaltene		190
	6.6.15	Corrosion		191
	6.6.16	Internal Corrosion Prevention		191
		6.6.16.1	Internal Coating	191
		6.6.16.2	Internal Corrosion Inhibitors	191
	6.6.17	External Corrosion Prevention		192
		6.6.17.1	External Coating	192
	6.6.18	Scales		192
	6.6.19	Scale Management		192
		6.6.19.1	Scale Inhibitors	192
	6.6.20	Erosion		193
	6.6.21	Mitigation Methods		193
		6.6.21.1	Reduction in Production Rate	193
		6.6.21.2	Design of Pipe System	193
		6.6.21.3	Increasing Wall Thickness	193
		6.6.21.4	Erosion-Resistant Material	193
6.7	Emerging Deepwater Technologies			193
	6.7.1	Dry Tree Semi-Submersibles		193
	6.7.2	Hybrid Riser System		195
	6.7.3	Free-Standing Flexible Riser System (FSFR)		196
	6.7.4	Multi-Lines Free Standing Riser		196
	6.7.5	Deep Steep Riser		197
	6.7.6	Expandable Monobore Liner Extension		197
	6.7.7	Smart Well Technology		199
	6.7.8	Autonomous Underwater Vehicles (AUVs)		200
	6.7.9	Nomad Systems		201
	6.7.10	Subsea Multiphase Pumps		202
	6.7.11	Subsea Processing		202

Contents   xiii

    6.7.12 Seabed Separation 203
    6.7.13 Subsea Pressure Boosting 204
  References 205

Chapter 7 Offshore Field Development 207

  7.1 Introduction 207
  7.2 Offshore Marginal Field Development Exercise 207
    7.2.1 Development Example.. 207
      7.2.1.1 Field Development Scenarios: Options/
        Alternatives.. 208
    7.2.2 Offshore Giant Field Development Exercise............ 209
      7.2.2.1 The Salient Data Is 209
      7.2.2.2 Solution Approach 210

Chapter 8 Health, Safety, and Environment in Offshore 213

  8.1 Basic Definitions... 213
  8.2 Human Factors.. 213
  8.3 Hazards on Oil and Gas Installations.............................. 215
  8.4 Procedural Aspects Related to Safety 216
    8.4.1 System Safety 216
      8.4.1.1 Process Safety and Hydrocarbon Release ... 216
    8.4.2 Process Leaks... 216
    8.4.3 Riser Leaks.. 216
    8.4.4 Fire and Gas Detection and Safety System 217
      8.4.4.1 Fire Protection 217
      8.4.4.2 Fire... 217
    8.4.5 Safety in Logistics Operations Related to
      Offshore Installation 220
      8.4.5.1 Vessel Collisions................................... 221
      8.4.5.2 Helicopter Incidents............................... 221
      8.4.5.3 Dropped Object..................................... 221
    8.4.6 Evacuation, Escape, and Rescue (EER) 222
      8.4.6.1 Reasons for EER.................................... 222
      8.4.6.2 Evacuation Sequence 223
      8.4.6.3 Life-Saving Equipment in EER................ 223
  8.5 Navigation Aids.. 225
    8.5.1 Emergency Position-Indicating Radio Beacon
      (EPIRB)/Search and Rescue Transponder (SART) ... 226
    8.5.2 Pyrotechnics 227
  8.6 Well Integrity... 228
    8.6.1 Well Failure: Example 1 229
    8.6.2 Well Failure: Example 2.. 229
  8.7 Life Extension and Assessment of Offshore Structures........ 230
    8.7.1 Structural Collapse.. 230
  8.8 Assessment Process 231

	8.8.1	Quality of Data...232
	8.8.2	Proof of Structural Integrity with Increased Loads....232
	8.8.3	Capacity and Performance of Damaged Structures..233
	8.8.4	Extended Life 233
8.9	IMO Resolutions...234
	8.9.1	Resolution A. 621(15) 234
	8.9.2	Resolution A. 671(16) 234
8.10	Offshore Facilities Interference..235
	8.10.1	Protection of Offshore Facilities/Rigs.....................235
8.11	Costliest and Deadliest Events in Oil and Gas Industry 236
	8.11.1	Bohai 2 Oil Rig Disaster, China (1979)...................236
	8.11.2	Alexander L. Kielland, North Sea, Norway (1980) ... 236
	8.11.3	Ocean Ranger Oil Rig Disaster, Canada (1982).......236
	8.11.4	Glomar Java Sea Drillship Disaster, South China Sea (1983) 237
	8.11.5	Enchova Central Platform Disaster, Brazil (1984)...237
	8.11.6	Piper Alpha, North Sea, United Kingdom (1988) 238
	8.11.7	Seacrest Drillship Disaster, South China Sea, Thailand (1989) 239
	8.11.8	Mumbai High North Disaster, Indian Ocean (2005)...239
	8.11.9	Usumacinta Jackup Disaster, Gulf of Mexico (2007) 240
	8.11.10	Deepwater Horizon, Gulf of Mexico (2010).............240
8.12	Offshore Security Threats 240
	8.12.1	Piracy..241
	8.12.2	Terrorism 241
	8.12.3	Insurgency 241
	8.12.4	Organized Crime...242
	8.12.5	Civil Protest..242
	8.12.6	Interstate Hostilities 243
	8.12.7	Vandalism..243
	8.12.8	Internal Sabotage...243
8.13	Trainings..244
8.14	Summary 244
References 244

Chapter 9 Legislations and Regulations in Offshore Operations around the World..247

9.1	Introduction 247
9.2	Europe 248
9.3	Norway 252
	9.3.1	Applicable Legislation..252
9.4	Kuwait..253
	9.4.1	Article 1..253

	9.4.2	Article 2	254
	9.4.3	Article 3	254
	9.4.4	Article 4	254
	9.4.5	Article 5	254
	9.4.6	Article 6	254
	9.4.7	Article 7	255
	9.4.8	Article 8	256
	9.4.9	Article 9	256
	9.4.10	Article 10	256
	9.4.11	Article 11	256
	9.4.12	Article 12	257
9.5		Australia	257
9.6		Egypt	260
	9.6.1	Regulation	260
	9.6.2	Directive on Offshore Safety	261
9.7		Qatar	261
	9.7.1	Government Policy Objectives	262
	9.7.2	Regulation	262
	9.7.3	The Regulatory Regime	262
9.8		Russia	263
	9.8.1	Regulatory Bodies	263
		9.8.1.1 Oil and Natural Gas	263
	9.8.2	Russian Legislation Regulating Foreign Investments	265
	9.8.3	Rules for Offshore Companies	265
9.9		India	266
	9.9.1	Domestic Production	266
	9.9.2	Government Policy Objectives	267
	9.9.3	Regulation	267
	9.9.4	Legal Framework on Minerals Mining in India	267
	9.9.5	Offshore Areas Minerals (Development & Regulation) Act, 2002	268
	9.9.6	Offshore Areas Mineral Concession Rules, 2006	268
	9.9.7	Other Regulatory Requirements	268
9.10		United States	269
	9.10.1	Government Policy Objectives	269
	9.10.2	Regulation	269
	9.10.3	Lease/License/Concession Term	270
9.11		Canada	270
	9.11.1	Regulation	271
9.12		Saudi Arabia	272
		References	273
Index			277

Foreword

In 1980, during my nascent days at the Indian School of Mines, as an undergraduate, I was trying to understand the nuances of petroleum technology at my alma matter, ISM Dhanbad. As a student of petroleum engineering, the very mention of the topic of offshore operations and engineering fascinated me and would trigger my imagination.

Offshore oil and gas technology was evolving in India during those days and operations had just begun. I vividly remember the news of oil discovery in Mumbai high offshore. There were no such books or literature in those days which could throw light on this subject for students like me. Till date also, there aren't many books in India on this topic. The authors of this book have designed the topics in a very crisp manner and in a fashion which amalgamates industry–academia requirement of the present times. This book will go a long way in igniting the minds of young people seeking deeper knowledge of offshore petroleum industry.

I happened to work in Mumbai offshore as a production engineer at BHN, which was the jewel of ONGC Mumbai high offshore operations during the nineties. With my hands-on experience, while working on offshore platforms, I can vouch that the author has done justice while selecting the topics in this book, and almost all critical aspects of offshore operations are covered in the best possible manner.

Today, as I go through the pages of this book, it reminds me of my graduation days and the desire to have a similar kind of book in those days. Offshore Operations and Engineering has been written primarily for aspiring energy professionals as well as experienced energy and management professionals who look forward to working in the offshore environment. The book encompasses topics particularly related to production and drilling operations related to offshore along with HSE, subsea engineering, ocean engineering, offshore logistics, safety and major legislations, as well as regulations governing international waters. It is a useful ready reckoner book and can be used as a curriculum book for undergraduate and postgraduate students in these subjects.

I expect this book will enrich budding engineers of petroleum industry with the first insight into offshore technology and operations, and to be of great value to aspiring energy professionals, existing engineers, and management personnel in the oil and gas sector.

Debasis Basu
Chairman
SPE India Section
Oil and Natural Gas Corporation Limited
Gujarat, India

Preface

The first commercial onshore oil production in 1859 in Pennsylvania revolutionized the energy sector of the world and ignited the growth of various global industries. Today's world is dependent on hydrocarbons to such a large extent that nearly 80% of the global primary energy needs are fulfilled by the hydrocarbon sector. Technological advancement coupled with innovations in the hydrocarbon sector provided the impetus for the industry to take a lead over other alternative energy resources sector. Thus, more and more onshore oil and gas is being produced.

Oil industry, however, is currently facing the biggest challenge to increase hydrocarbon production as the gap between production and consumption has been increasing since 1950 and is likely to continue. At the same time, easy oil has already been produced. We need to move to difficult terrains for continued exploration and production. To bridge the gap between consumption and production, industry has taken a leap by moving from shallow water to deep water, and now aims to move further to ultradeep water. However, offshore operations are highly challenging compared to onshore. A typical offshore platform needs to be self-sufficient in regards to energy and water requirements, housing electrical generation, water desalination, as well as all the equipment necessary to proceed with drilling and production operations. In addition to electricity generation, another important aspect is the use of sea water and its conversion into potable water using water makers.

Owing to the harsh environmental conditions offshore, operations face many challenges compared to onshore operations. Some of these challenges include corrosive environment, logistics of transporting man and machine along with accommodation for manpower due to distance from the shore, upkeep and maintenance of system and equipment, power generation, evacuation of crude oil (from well and to shore with related flow assurance issues), etc. Along with operations, paramount importance should be given to the health and safety of the crew and keeping the surrounding marine environment pollution free.

This book provides a comprehensive understanding of each and every aspect of offshore operations taking into consideration all the conventional methods of operations, emerging technologies, legislations, as well as the health, safety, and environment impact of offshore operations.

The book begins by providing an overview of notable offshore fields across the globe and the statistics of present oil production. It provides detailed descriptions of all types of platforms available for offshore operations along with their structural details. The book presents a comprehensive analysis of the different available fixed and floating platforms. It also describes mooring and anchoring systems utilized to provide the station-keeping of floating offshore structures. Furthermore, riser systems used to recover hydrocarbons are explained. Being one of the key elements of offshore operations, offshore drilling and completion (planning and designing) have been elaborated.

After drilling and completion, production, storage, and transportation are the concluding part of onshore as well as offshore operations. Thus, the book discusses production equipment, safety systems, automation, storage facilities, and transportation in detail. New technologies and innovation will be the key to drive toward deep water where the future of oil and gas lies. Therefore, emerging technologies likely to support drilling and production in deepwater along with engineering intricacies of subsea completion has been separately discussed in the book. In addition, legislations play a major role for every country to protect their environment and ensure the proper use of the oil and gas treasure. To provide better understanding of legislations, the book ends with common legislation acts and a comparison of different legislation acts from various major oil and gas producing nations.

Acknowledgments

The authors are grateful to their families for their invaluable support, without which this book would not have been possible. We want to thank Pandit Deendayal Petroleum University, Gandhi Nagar, for providing all the support for publishing this book.

We appreciate the contributions of upstream undergraduate students (Class of 2018) for discussing topics related to platform operations. Further, we appreciate Mr. Shivshambhu Kumar and Mr. Vivek Thakkar for proofreading the draft versions and preparing the list of references. The book would have remained incomplete without the invaluable feedback of the reviewers, and we are grateful to them for their input.

We would also like to show our immense gratitude toward the editors and the publisher (Taylor & Francis) for accepting our book for publication. Finally, we would like to acknowledge all the authors and contributors of the different articles, research papers, and online sources we have used in our book.

Authors

Shashi Shekhar Prasad Singh graduated in Petroleum Engineering from IIT (ISM) Dhanbad in 1969, and has served the National Oil Company of India, ONGC, for nearly four decades at senior positions in its operations and R&D activities. At the Institute of Oil and Gas Production technology (IOGPT), an R&D wing of ONGC, he was responsible for the identification and assimilation of prospective new technologies for offshore and played a key role in deepwater development ventures in Indian offshore. As the head of the Offshore Development Group at IOGPT in the late 1990s, he introduced innovative ideas and cost-cutting technologies which helped to develop many small and marginal offshore fields at later dates. He presented a status paper on oil and gas scenario in Indian deepwater in Deep Tech-97 (an international conference organized under the aegis of Ministry of Petroleum and Natural Gas, Government of India). It helped to bridge the technology gap and mobilize huge resources required for deepwater development in India. He has been associated with Pandit Deendayal Petroleum University since 2008 as an adjunct professor and has been teaching offshore operations along with other petroleum engineering courses.

Jatin R. Agarwal earned a PhD under the guidance of Professor Subhash N. Shah, Emeritus Professor, Oklahoma University and a Shell Total Chair Professor at PDPU in the area of production enhancement. Dr. Agarwal also holds a bachelor's degree in chemical engineering from VNSGU and two master's degrees in petroleum engineering from PDPU and the University of Tulsa. He has 9 years of research experience in flow assurance and artificial lift systems. He has published several papers in different journals and at conferences on topics related to flow assurance for onshore and offshore. Currently, he is working as the Manager of the School of Petroleum Technology at PDPU, and is involved in both teaching and teaching-related administrative jobs at PDPU. His teaching profile includes introduction to petroleum engineering, pipeline engineering, offshore operations, and advanced drilling technology at the undergraduate level. Presently, he has established a state-of-the-art Drilling Stimulation and Cementation Research Center that caters to both offshore and onshore needs in

India. The research center focuses on designing high-temperature and high-pressure drilling fluids, cement slurries, and frac fluids for offshore operations. In addition, the center has expanded its horizon for the study of enhanced oil recovery for both onshore and offshore fields. Currently, he is involved in consultancy work provided by ONGC, GSPC, JTI, GNRL, HOEC, OIL, and many other oil and gas companies. He has successfully delivered a scale-up theory for predicting wax deposition in offshore operations.

Nag Mani is an alumni of BIT Mesra, Electrical Engineering Department. Over a career spanning 35 years in upstream hydrocarbon sector, he has worked with Oil and Natural Gas Corporation (ONGC), Reliance Industries Ltd (RIL) and Gujarat State Petroleum Corporation (GSPC). At ONGC, he worked in onshore and offshore divisions like SHP, which is one of the biggest Offshore platform of ONGC. He started working with RIL in Subsea division of the KGD6 Project, the first Deep sea project in India and a leading field in India. His responsibilities included execution of different type of Subsea jobs ranging from Installation of equipment like Subsea X-Mass tree from MSV to Inspection, Maintenance and Repair/Replacement jobs in Subsea. With GSPC, he worked with Offshore Projects like Deen Dayal field in East Coast of India for Subsea Inspection, Maintenance and Repair Division with additional responsibility of debottlenecking works for Offshore and Onshore Operation. Deen Dayal field is a Sour gas field with H_2S and Co_2 presence in gas.

1 Introduction to Offshore Operation

1.1 OCEAN BASELINE

Oceans cover almost three-fourths of the earth's surface, and the land beneath them can meet the energy requirements of the world for years to come. Beaches extend from the shore into the ocean on a continental shelf that gradually descends to a sharp drop called the continental slope. The continental shelf can be as narrow as 20 km or as wide as 400 km. Water on the continental shelf is usually shallow and rarely more than 150–200 m deep. Continental shelf drops off at the continental slope, ending in abyssal plains up to 3–5 km below the sea level. While many plains are flat, others have jagged mountain ridges, deep canyons, and valleys. The tops of some of these mountain ridges form islands where they extend above the water. Our borders extend 200 miles into the water from coastlines, encompassing areas larger than some countries. This large underwater area is called the exclusive economic zone (EEZ).

Territorial sea baseline is defined by the United Nations (UN) as "the line from which the seaward limits of a State's territorial sea and certain other maritime zones of jurisdiction are measured". These zones include the breadth of the territorial sea; the seaward limits of the contiguous zone, the EEZ; and, in some cases, the continental shelf. The territorial sea baseline varies according to the shape of the coastline as follows:

- The normal baseline corresponds with the low-water line along the coast, including the coasts of islands. According to the Convention, a normal baseline can be drawn around low-tide elevations defined as naturally formed areas of land surrounded by and above water at low tide but submerged at high tide, provided they are wholly or partly within 12 nautical miles of the coast [1].
- Straight baselines are straight lines joining specified or discrete points on the low-water line, usually known as straight baseline endpoints. These may be used in localities where the coastline is deeply indented and cut into or where there are a fringe of islands along the coast in the immediate vicinity [1].
- Bay or river closing lines are straight lines drawn between the respective low-water marks of the natural entrance points of bays or rivers.

Territorial sea, being an integral part of the continental shelf, is an important aspect to consider.

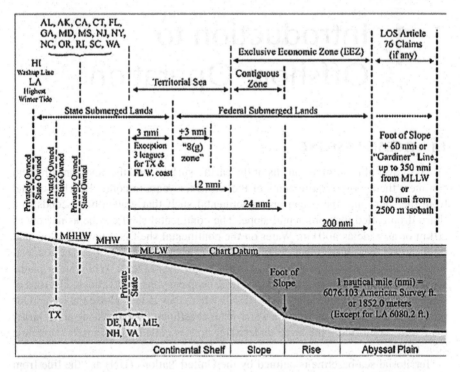

FIGURE 1.1 Water according to the United Nations Convention on the Law of the Sea, 1982 [2].

Baselines are the starting point from which the territorial sea and other maritime zones of jurisdiction are measured, for example, contiguous zone, EEZ, and continental shelf (Figure 1.1).

1.2 OCEAN ENVIRONMENT

The ocean environment greatly influences the structural design, logistics, and operational safety.

1.2.1 Metocean Condition and Study

Environmental conditions/metocean conditions have a direct bearing on any coastal or offshore project, as well as the project's operation and maintenance. The selection of equipment, system, location, operational strategy, etc. also depends on these conditions. In addition, environmental conditions influence the financial decision-making regarding the project and operation.

It is essential to understand the environmental/metocean conditions by studying all the parameters, including waves, wind, current, seasonal variation, and probability of cyclones and storms. These studies may be conducted as per the requirements of the job (Figure 1.2).

Introduction to Offshore Operation

FIGURE 1.2 Data well waverider buoy deployed near the southwestern coast of France to measure ocean wave statistics, including significant wave height and period, wave direction, and power spectrum [3].

1.2.2 Meteorology

- Wind analysis including speed, direction, gustiness, seasonal variance, and spectrum.
- Air temperature profiling including seasonal variance, humidity, and precipitation.
- Sea water temperature profiling from mean sea level to sea bottom including seasonal variance.
- Sea current analysis.
- Adverse weather analysis including typhoons, hurricanes, cyclones, and tsunamis.

1.2.3 Physical Oceanography

- Sea water level changes including historical, expected, and seasonal tides and wave and storm surge analysis.
- Seiches.
- Wind waves – wind seas and swells – characterized by significant wave heights and periods, as well as propagation direction and spectra (Figure 1.3).
- Bathymetry.
- Analysis of salinity, sea water temperature at different depths, and other parameters.
- Analysis of ice occurrence including extent, thickness, and strength.

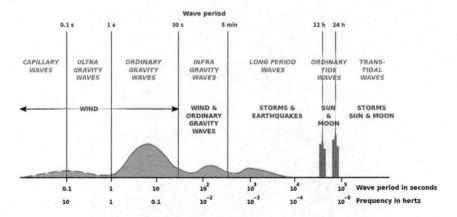

FIGURE 1.3 Classification of the wave phenomena – of the sea and ocean surface – according to wave period by Walter Munk [4].

1.2.4 Metocean Data

Metrology and physical oceanography provides the basis for understanding metocean conditions. Metocean data are collected by deploying measuring instruments, through satellites by remote sensing, and by remodeling of existing models including latest inputs. Metocean data form the basis of metocean conditions (Figure 1.4).

1.3 OFFSHORE OIL AND GAS PRODUCTION

To meet the ever-increasing market demand for oil and gas, offshore prospecting and production of oil and gas, apart from onshore production, have become essential.

FIGURE 1.4 Ice beacon – for tracking ice movement by GPS, as well as other sensors for measuring more metocean parameters – and Pablo Clemente-Colón of the U.S. National Ice Center [3,5].

Introduction to Offshore Operation

Technological advancements have helped in reaching deepwater sources of oil and gas.

1.3.1 Comparison of Onshore versus Offshore Oil Production

It is clear from Figures 1.5 and 1.6 that, out of the total oil production, 67% is onshore production, and only 33% is offshore production, out of which 24% is from shallow water. Although since 2000 production from deepwater has continuously increased due to technological advancements, deepwater contributes to only 9% of the total production. Thus, to bridge the gap between production and consumption, we have to produce more oil from deep and ultradeep water (Figures 1.5 and 1.6).

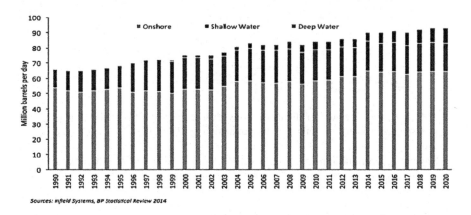

FIGURE 1.5 Statistical analysis of onshore and offshore oil and gas.

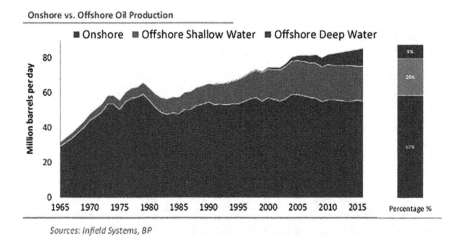

FIGURE 1.6 Comparison of total oil production from onshore and offshore fields since 1965.

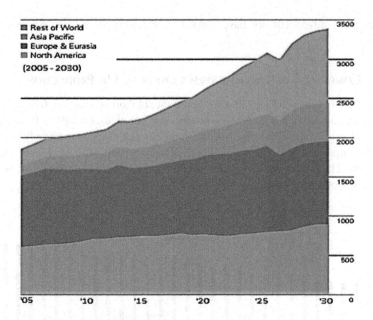

FIGURE 1.7 Oil forecasts explicitly showing that deepwater oil production will have a similar weightage as that of shallow water production by 2020 [6].

Natural gas production increased by 1.1% from 2016 to 2017, and is expected to increase by approximately 4.3% by 2030 (Figure 1.7).

1.3.2 COMPARISION OF OFFSHORE OIL PRODUCTION ON THE BASIS OF WATER DEPTH [7]

See Figures 1.8–1.10.

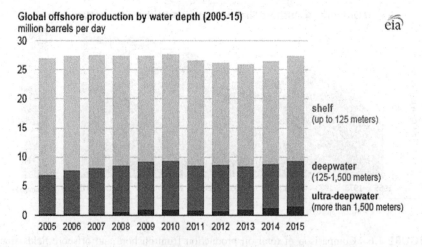

FIGURE 1.8 Offshore production by water depth – 27% from deepwater and 73% from shelf.

Introduction to Offshore Operation

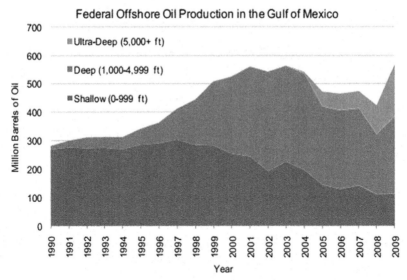

FIGURE 1.9 Offshore oil production according to water depth for different regions. In Brazil and United States, production from ultradeep has increased noticeably from 2005 to 2015 [7].

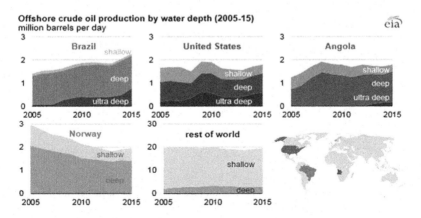

FIGURE 1.10 Federal offshore oil production especially in the Gulf of Mexico [7].

1.3.3 Rig Count and Utilization Rate [8]

See Figures 1.11 and 1.12.

1.4 OFFSHORE OPERATIONS

The first well drilled over water was completed in 1897. However, real offshore oil exploration began in the late 1930s, and the first platform was installed in the Gulf of Mexico in 1945. Initially, wells were drilled from wooden jetties or piers. However,

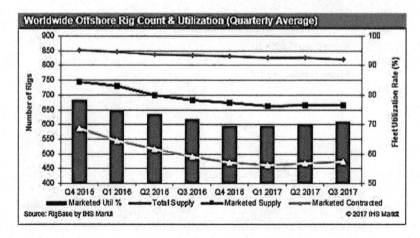

FIGURE 1.11 Worldwide distribution of oil rigs. Gulf of Mexico has the highest oil rigs, followed by North Sea, Southeast Asia, Far East Asia, and Persian Gulf.

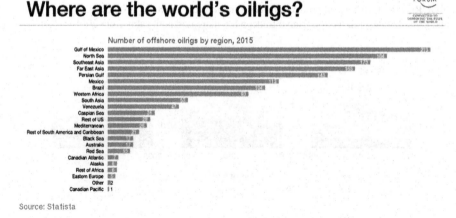

FIGURE 1.12 Number of rigs and utilization rate worldwide. Worldwide total supply has decreased from 740 in 2015 to 660 in 2017 [9].

with technical advancements, these wooden structures were replaced by steel structures. Till then these structures were fixed and immobile. For ease of movement and better flexibility, mobile units were introduced in early 1950s. Subsequently, to accommodate increased payload, water depth, and rough sea conditions, concrete gravity structures were developed. With a shift toward deepwater exploration, subsea production systems were developed.

While bottom-supported units were being developed for shallow waters, floating/buoyant vessels were being developed for water depths beyond 500 ft. The first floating drilling vessel was commissioned in 1953, which was capable of drilling at depths ranging from 400 to 3,000 ft. The Glomar Challenger was the first versatile deepwater drillship with dynamic positioning capabilities to hold station at a water depth of 20,000 ft and drilling depth of 25,000 ft equipped with satellite navigation facilities (Global Marine Co., USA, commissioned in 1968).

The semi-submersibles were developed by adding buoyant hulls to help drill while afloat and not sitting at the mud line. These are most suitable for extreme sea conditions prevalent in the North Sea. In addition to developments in mobile units, platform technology saw significant development. The first platform was installed off the coast of Louisiana at a water depth of 20 ft in 1947. The first platform at a water depth of 100 ft was installed in 1955 and at 200 ft in 1959. During 1970s, the North Sea was the focal point of offshore activities, and due to heavy load requirements, the attention was shifted toward structures. The different types of structures available include template/jacket, guyed towers, tension leg platform, concrete gravity platform, and tripod tower platform.

The first underwater completion was done in Lake Erie in 1959 by Placid Oil & Gas Co. in view of boat traffic and icy conditions during winter. The first underwater completion in open sea was done in 1960 off the coast of Peru by Peruvian Pacific Oil Company.

In December 1960, Shell Oil Company completed an underwater well (subsea) in Louisiana after 7 years of experimental design and evaluation. The techniques, methods, and equipment for completing, producing, and working over wells have advanced tremendously since the first subsea well was completed in the late 1950. Complex multiwell systems have been installed on the seabed. Nowadays, in view of the hydraulic, electronic, and acoustic signal exploratory activities in the deeper continental shelf, there is a tremendous scope to improve the technology. With technological advancement, offshore oil and gas industry can explore and exploit deep and ultradeep locations. Figure 1.13 depicts the current status of the oil and gas industry.

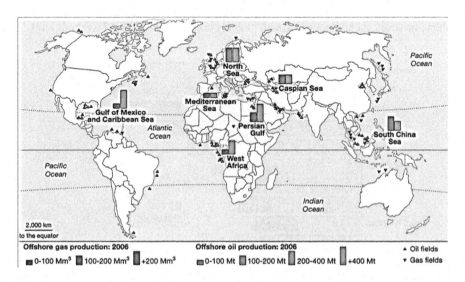

FIGURE 1.13 World map indicating major offshore fields along with their oil and gas potential [10].

1.4.1 Notable Offshore Fields across the Globe

Some of the notable offshore fields are [11]

- The North Sea
- The Gulf of Mexico (offshore Texas, Louisiana, Mississippi, and Alabama)
- California (in the Los Angeles Basin and Santa Barbara Channel, Ventura basin)
- The Caspian Sea (notably some major fields offshore Azerbaijan)
- The Campos and Santos Basins off the coast of Brazil
- Newfoundland and Nova Scotia (Atlantic Canada)
- Several fields off West Africa most notably west of Nigeria and Angola
- Offshore fields in Southeast Asia and Sakhalin, Russia
- The Persian Gulf including Safaniya, Manifa, and Marjan belonging to Saudi Arabia
- India (Mumbai High, K G Basin, East Coast of India, Tapti Field, Gujarat)
- The Taranaki Basin in New Zealand
- The Kara Sea, north of Siberia

1.4.2 Major Offshore Oil Producing Countries

According to the analysis, Saudi Arabia will continue to be the number one offshore producer with an estimated 4.8 MMb/d, of which 3.9 MMb/d is liquid production. The Safaniya, Manifa, and Qatif oil fields, as well as Berri, will continue to contribute the most in terms of production. Norway is estimated to produce just over 4 MMb/d, of which approximately 50% is gas production, most noticeably from Troll and Ormen Lange. The gas-dominant countries of Qatar and Iran are expected to produce 3.9 and 3.3 MMb/d of expected total offshore production, respectively. Iran and Saudi Arabia show the strongest production growth globally compared to 2016 [12] (Figure 1.14).

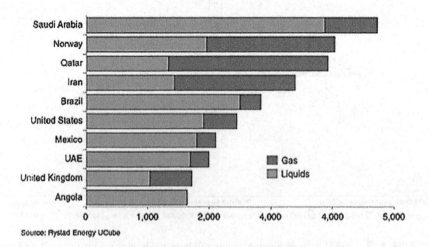

FIGURE 1.14 Top 10 offshore producing countries in 2017 (MMb/d) [10].

Introduction to Offshore Operation

1.4.3 OFFSHORE FACILITIES (INSTALLATIONS/PLATFORMS/RIGS)

An offshore installation can be an offshore process platform or an offshore drilling rig. Normally, an offshore process platform is a structure as well as a facility to process and transport petroleum and natural gas produced from a producing well. Offshore drilling rig is a structure as well as a facility to drill a wellbore below the seabed to explore and extract petroleum and natural gas present in rock formations beneath the seabed [12]. Depending on the situation, the offshore platform may have drilling rigs installed on the same structure. Modular drilling rigs are also installed on offshore platforms. Otherwise, standalone offshore process platforms and offshore drilling rigs are deployed. Offshore facilities may be manned or unmanned, with corresponding facilities depending on the work requirement. Unmanned locations are operated remotely.

1.4.4 TYPES OF OFFSHORE INSTALLATIONS/PLATFORMS

- Fixed platform
- Compliant tower
- Tension leg platform
- Floating production unit (Conventional)
- Truss spar
- MiniDOC spar
- Classic spar
- Cell spar

1.4.5 TYPE OF OFFSHORE RIGS

- Jackup rigs
- Semi-submersible rigs
- Drillships

1.4.6 CHALLENGES WITH OFFSHORE FACILITIES

- Remote location and adverse environment make offshore oil and gas operation much more difficult and challenging than onshore operation. Several new and innovative technologies are being used to meet the challenges associated with offshore operations.
- Well depth along with ocean depth adds up to significant hydrostatic head for fluid circulation in case of drilling operation, as well as for lifting production fluid, which in turn requires large amounts of power.
- Transfer of man and material and accommodation of manpower are also challenges associated with offshore production.
- Safe working/living conditions need to be ensured even during extreme weather and environmental scenarios. The asset and equipment above and below the water level need to be kept in safe working conditions, which is a significant challenge due to limited logistical support.

12 Offshore Operations and Engineering

- Oil prices are highly volatile and susceptible to large price drops as observed recently due to world geopolitics. Compared with onshore energy projects, offshore oil and gas projects involve heavy investments and are much more sensitive to price variability. Hence, offshore energy production needs to optimize its production and operation costs [13]. This can be achieved by deploying and utilizing appropriate new technologies and systems without compromising safety, efficiency, operability, maintainability, and reliability.
- Accidents are very difficult to deal with in offshore oil and gas industries due to logistical constraints. Fire accidents are arguably one of the most dangerous incidents and the most feared by the offshore crew. Ensuring adequate safety equipment and detection systems is essential for all offshore installations.
- Communication is a critical aspect of business operations.

1.4.7 ECOLOGICAL AND ENVIRONMENTAL RISKS

Extraction of hydrocarbon involves dealing with volatile substances under varying temperature and pressure conditions, sometimes even under extreme pressure and temperature. Adverse/hostile environmental conditions may result in unfavorable events such as accidents and loss of human life. Therefore, the main aim should be to protect human life and the surrounding environment from contamination by the extracted fluids. Some of the biggest oil platform accidents include:

1. Piper Alpha (167 fatalities). Cause: Condensate leakage
2. Alexander Kielland (123 fatalities). Cause: Harsh weather/strong winds
3. Enchova Central Platform (42 deaths). Cause: Blowout, fire, and explosion
4. Deepwater Horizon (11 deaths). Cause: Gas kick, leading to fire
5. Mumbai High North (22 deaths). Cause: collision with Multi Support Vessel (MSV) and subsequent fire after gas leakage [14].

Marine life is also significantly affected by offshore facilities/platforms. Offshore hydrocarbon operations are sometimes carried out near water sport and commercial fishing destinations, such as the Gulf of Mexico, which requires additional safety and security measures to ensure trouble-free operations.

Rigs to Reefs program is currently being actively pursued by authorities in the United States and Brunei. Decommissioned installations such as platforms or rigs are either left in place in sea or towed to a new location to become a permanent artificial reef.

In addition to offshore fires, oil spills and pipeline ruptures or leaks during transportation of liquid hydrocarbon from offshore to onshore facilities by oil tankers or pipelines or due to problems on platforms or rigs are major environmental risks in offshore operations.

Water is also produced with hydrocarbons and brought to the surface. This water is called produced water, which may be highly saline with possible presence of hydrocarbons.

Introduction to Offshore Operation

1.5 HOW ARE OFFSHORE OPERATIONS DIFFERENT FROM ONSHORE

The basic difference between offshore and onshore installation is the remote and harsh locations of offshore installations.

1. Offshore installations are mostly self-sufficient in terms of power, water, and processing facilities. On the other hand, onshore installation need not be self-sufficient in terms of power and water.
2. Logistical support to offshore installations is limited to air and marine transportation, which is dependent on weather conditions and may hinder continuous approach. On the other hand, logistical support to onshore installations can be provided anytime without any constraints.
3. On offshore installations there is limited availability for manpower accommodation, which is not a concern on onshore installations.
4. Processes and equipment are selected to meet the harsh and corrosive environment of offshore operations.

1.5.1 SAFETY REQUIREMENTS

Safety adherence and management is a requirement of the petroleum industry. In offshore sector, safety becomes more demanding due to remote and adverse weather conditions. Production and processing of oil and gas involves handling volatile substances at varying temperatures and pressures. Any accident in such a situation can turn into a disaster, resulting in loss of human life and assets, as well as damage to the environment. Protection of human life should always be the priority.

Proper training with awareness and alertness are key to avoid and minimize any such unfortunate accidents. All companies operating offshore have safety management systems to cover their entire operation.

1.6 TECHNOLOGY WAVE

Oil industry is currently facing the biggest challenge to increase well productivity as oil consumption is increasing day by day, particularly in the developed countries and developing countries. Till now, we have produced most of the crude oil from shallow water. To bridge the gap between consumption and production, we need to explore deep sea and ultradeep sea, necessitating technological advancements.

It took the petroleum industry 50 years of inching up the offshore learning curve before they placed the historical creole platform at 30 ft in the Gulf of Mexico water in the 1947. In the next 50 years, remarkable technologies were developed with the industry soaring nearly to the top of the curve.

There is an increasing need to reduce the cost for the economical monetization of offshore marginal fields in India. In recent years, there has been remarkable development in the technology for marginal offshore fields. Some of the prospective new technologies/innovations at various stages of implementation are discussed below.

To meet the ever-growing demand of hydrocarbons, we are left with no alternative but to move to offshore operations (shallow to deep and deep to ultradeep).

14 Offshore Operations and Engineering

To achieve this goal many joint research projects (JRP) are currently ongoing to develop suitable technologies, some of which are listed below:

- Riserless drilling
- Intelligent wells
- Subsea processing
- Dual activity drilling rigs
- Subsea pumping
- Surface blowout preventer

Although exploration and exploitation of oil and gas from offshore fields have always remained cost and time intensive, it has not stopped companies and countries from operating in offshore fields – shallow, deep, and deepwater.

Advancement in technologies along with maturing technology, not only in the field of exploitation but even in exploration, has resulted in making offshore operations very interesting and lucrative, resulting in better economy for offshore. New frontiers and fields such as Guyana or Deepwater Outeniqua Basin Offshore South Africa are opening new avenues. Similarly, new finds in matured fields in UK North Sea, such as Glengorm, have brought back offshore operations in focus.

REFERENCES

1. United Nations Convention on the Law of the Sea, United Nations, New York, 1973. United Nations, Baselines: National Legislation with Illustrative Maps, United Nations, New York, 1989, Part 2 Territorial Sea and Contiguous.
2. United Nations Convention on the Law of the Sea, United Nations, New York, 1973. United Nations, Baselines: National Legislation with Illustrative Maps, United Nations, New York, 1989, Part 1: Introduction.
3. *Metocean.* Available at https://en.wikipedia.org/w/index.php?title=Metocean&oldid= 856824812.
4. W.H. Munk, Origin and generation of waves, *Coast. Eng. Proc.* 1 (2017), p. 1.
5. *NOAA Photo Library.* Available at www.flickr.com/photos/noaaphotolib/5037012714/.
6. BP, 67th edition Contents is one of the most widely respected, Stat. Rev. World Energy (2018), pp. 1–56.
7. U.S.E.I. Administration, *Annual Energy Outlook 2015.*
8. I.P. Rigbase, *IHS Petrodata RigBase*, IHS Markit (2017).
9. *Number of offshore rigs worldwide.* Available at www.statista.com/statistics/279096/ number-of-offshore-rigs-worldwide-by-operator/.
10. Rystad Energy, *UcubeFree Rystad Energy*, Rystad Energy (2017).
11. *Oil platform.* Available at https://en.wikipedia.org/w/index.php?title=Oil_platform& oldid=886522136.
12. *Oil platform wikivisually.* Available at https://wikivisually.com/wiki/Oil_platform.
13. *Most current and crucial challenges confronting the onshore and offshore oil and gas pipeline industry today?* Available at www.quora.com/What-are-the-most-current-and-crucial-challenges-confronting-the-onshore-and-offshore-oil-and-gas-pipeline-industry-today.
14. *Deepwater Horizon explosion.* Available at https://en.wikipedia.org/wiki/Deepwater_ Horizon_explosion.

2 Offshore Structure and Design

To facilitate drilling or production operations at sea the first and foremost requirement is to locate suitable spaces to install the required equipment as well as to accommodate the operating crew with requisite facilities to ensure round-the-clock operations. Although certain utilities such as potable water and electricity are not readily available at these sites, they are required for operations, as well as for the operating crew, their living accommodation, and the associated basic needs. In fact, operations of this magnitude need an artificial island called offshore facilities/platforms. Exploration and production companies use fixed structures up to the threshold of deepwater for drilling and the subsequent production of oil and gas. Deep and ultradeep locations use buoyant/floating facilities to support topside and downside connected project load, as well as logistical support required for marine operations. Therefore, offshore structures conventionally used for oil and gas drilling and production operations are classified as either fixed platforms or floating platforms (Figure 2.1).

This chapter discusses the different types of structures in the abovementioned categories along with their designs.

Fixed structures Fixed structures physically sit on the bottom of the sea and are held in place either by the sheer weight of the structure or by steel piles driven into

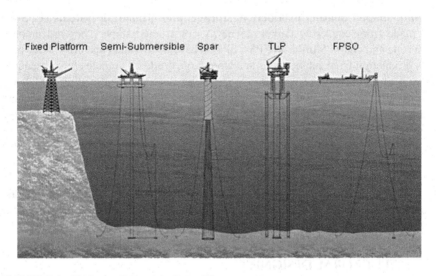

FIGURE 2.1 Various offshore platforms [1].

the seabed and affixed to the structure. The different types of fixed structures are discussed below.

Jacketed platforms consist of a jacket and a deck. The jacket is the tall vertical section built from tubular steel members and affixed to the seabed by driven piles. The deck placed on the top holds the production equipment and facilities on the top connecting subsea facilities/equipment including pipeline and risers.

Compliant towers are made of tubular steel members, similar to the fixed platform. These towers are fixed to the bottom with piling and support a deck with topsides.

Gravity platforms are built from reinforced concrete. With their substantial weight, as well as the weight of the topsides, these platforms rely on gravity to hold them in place.

Floating platform systems These include tension-leg platforms (TLPs) and moored floating systems of different shapes. These platforms need to be held in place with steel pipe tendons or moored in place with chains on the seabed attached using wire or polyester rope.

Spar platforms float as a result of deep large-diameter hollow cylinders that are weighted at the bottom to keep them upright. Wire or synthetic rope-and-chain combinations moor the hulls to the seabed.

Tension-leg platforms (TLPs) have floating hulls made of buoyant columns and pontoons. Steel pipe tendons hold the hulls down below their natural level of flotation, keeping the tendons in tension and hulls in place.

Floating production systems (FPSs) consist of ship-shaped or semi-submersible hulls with onboard production facilities. Wire or synthetic rope and chain moor them in place. Strong sea waves and wave current call for special equipment to accommodate risers that transport oil and gas from the sea floor wet trees to the production facilities on the deck.

Ship-shaped floating production, storage, and offloading systems (FPSOs) are made from converted tankers or newly constructed ships. They are moored with rope and chain. Similar to FPSs, FPSOs have no drilling capability. They process production from subsea wells and store large crude oil volumes for subsequent transport by shuttle tankers. A variation, floating storage and offloading (FSO) system, receives processed oil from nearby platform FPSs and stores it for subsequent transport by shuttle tankers. (These units are often referred to as floating storage units [FSU]). An FPS and FSO/FSU are collectively equivalent to an FPSO. Water depths present no limitation to FPSOs and FSOs. Cylindrical-hull FPSOs have double-bottom, double-sided hulls with enough ballast to maintain their position in safe conditions.

Floating drilling production storage and offloading systems (FDPSOs) are another variation of the FPSO. These can either be ship-shaped or cylindrical hulls, and accommodate onboard drilling operations [2].

2.1 STRUCTURAL DESIGNING

Offshore structures are continually exposed to harsh environmental conditions, and are affected by sea water, winds, waves, storms, earthquakes, and tsunamis.

Thus, various protective measures are required to keep these structures without any damage and strong. Of these measures, corrosion protection is the most critical.

2.1.1 Corrosion Protection

Corrosion is metal deterioration resulting from an electrochemical or chemical reaction in its surrounding environment.

An active corrosion cell primarily contains the following four components:

1. An anode (electrode from where electrons are emitted resulting in metal loss).
2. A cathode (electrode where electrons collect).
3. A metallic path (often the structure itself provides the metallic path).
4. An electrolyte is the medium in which the anode and cathode are immersed (the electrolyte could be any moist surface or immersion in any conducting fluid, water, or soil) (Figure 2.2).

2.1.2 Cathodic Protection

Cathodic protection is one of the most common corrosion protection techniques used today. Its concept and application are ingeniously simple, stemming from the basic theory of corrosion, the active corrosion cell. The presence of all four components allows the electrochemical reaction to proceed, resulting in electron flow from the anode to the cathode and causing corrosion of the anodic material. In cathodic

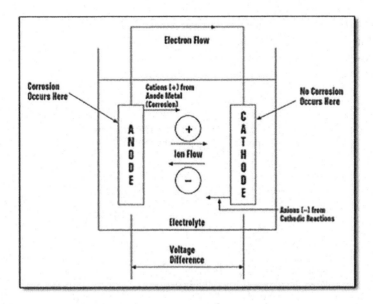

FIGURE 2.2 Corrosion cell explaining the relations of the four key components of an active cell [3].

protection, enough electrons are supplied to the anodic area that it becomes cathodic, thus slowing down or eliminating anode deterioration altogether. For this, an external anodic material is used that supplements electron supply to the area at risk of corrosion.

2.1.2.1 Impressed Current Cathodic Protection Systems

Cathodic protection is applied through one of the two principal methods: impressed current or sacrificial anodes. Impressed current differs from sacrificial anodes as it creates an electrochemical cell driven by an external power supply. The basic components required to achieve this circuit are a power source, an electrode, an electrolyte, and the structure requiring cathodic protection. Figure 2.3 illustrates the circuit components. The DC power source supplies the electrode with a positive current which is then transferred onto the structure ionically and returns back to the power supply. For power in offshore structures, alternative sources such as gas engines, renewable energy, or thermoelectric generators need to be utilized.

2.1.2.2 Sacrificial Anodes (Galvanic Action)

Unlike the impressed current method, the use of sacrificial anodes does not require an external power supply. Instead, it utilizes the natural tendency of electron flow from negative to positive potential to drive the current. This phenomenon is also referred to as galvanic action. As shown in Figure 2.4, electrons are supplied to the protected structure by the anode via an electrical connection. Anode reactions driven by the environment produce metal ions that act as positive current flowing toward the structure to complete the circuit. In practice, the anode is commonly mounted on a steel core and attached directly to the structure.

In addition to the abovementioned protections, various loads acting on the offshore structures are also taken into account.

Corrosion-resistant paint and coats are applied in the splash zone (transition area of jacket structure at the water entry point) to avoid corrosion.

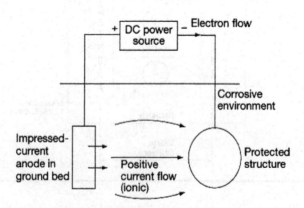

FIGURE 2.3 Typical components of an impressed current cathodic protection system [3].

Offshore Structure and Design

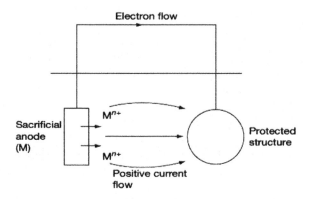

FIGURE 2.4 Typical components of a sacrificial anode [3].

2.1.3 Loads

It is essential to understand the load types that contribute to the design of offshore structures.

1. Constant loads
2. Variable loads
3. Environment-dependent loads
   - Wind and wave-induced loads
   - Earthquake loads
   - Ice and snow-induced loads
   - Temperature-induced loads
   - Marine growth
4. Additional loads during installation and construction
5. Unforeseen loads

Wind and wave-induced loading has a significant impact on the design of offshore structures.

2.1.3.1 Constant Loads

These are also known as dead or permanent loads, and constitute the weight of bare structures in air, ballast, equipment, and associated structures permanently mounted on the platform with appropriate adjustment for hydrostatic and hydrodynamic forces on the members below the waterline [4].

2.1.3.2 Variable Loads

These are also known as operating or live loads, and include the weight of non-permanent equipment or material or manpower. Other contributing factors are forces generated during drilling, production, helicopter operation, etc. BS 6235 recommends the following live load values:

Crew quarters and passageways: 3.2 KN/m^2
Working areas: 8.5 KN/m^2

2.1.3.3 Environment-Dependent Loads (Normally Calculated on Historically 100-Year Return Period)

2.1.3.3.1 Wind and Wave-Induced Load

Wind and wave-induced loads play a significant role in the design of offshore installations. Wind and waves jointly or individually exert large forces on offshore structures, threatening their stability. These forces introduce complexity in the design of the structure. Wind data is normally standardized at a specific height from the mean sea level for proper consideration.

The following loadings are considered most unfavorable according to some wind and wave codes.

- One-minute sustained wind speeds combined with extreme waves.
- Three-second gusts.

Waves driven by wind is the most crucial function in designing offshore platforms as they may approach platforms from more than one direction with irregular shapes and heights. These forces are determined by considering complex functions. Wave surface profiling and kinematics are useful in assessing such forces to a significant extent.

2.1.3.3.1.1 Wave Theories Wave kinematics are described by wave theories. These theories help calculate the particle velocities and accelerations and the dynamic pressure as functions of the surface elevation of the waves. Waves are assumed to be long-crested, that is, they can be described by a two-dimensional flow-field, and are characterized by wave height (H), period (T), and water depth (d) (Figure 2.5).

Morison's equation can be applied when $D/L < 0.2$, where D is the member diameter and L is the wavelength [4].

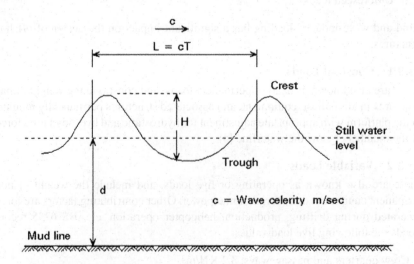

FIGURE 2.5 Wave symbols [5].

Offshore Structure and Design

Morison's equation expresses the wave force as the sum of:

- An inertia force proportional to particle acceleration.
- A non-linear drag force proportional to the square of particle velocity.
- $F(t) = F_I + F_D = \rho C_m V \dot{u} + \dfrac{1}{2} \rho \dot{C}_d A u |u|$

Where,
$F(t)$ = total inline force on the object
F_I = inertial force
F_D = drag force

2.1.3.3.2 Earthquake Load

Seismic study of the installation site should be conducted before designing offshore installations. Correlation of these results with historical data should also be done to adhere to the agreed design standards for mitigating the effects of earthquakes. Nowadays, most of the area worldwide is divided in different seismic zones for earthquake assessment. Seismic studies include investigating seafloor instability, scouring, etc.

Designing offshore structures capable of withstanding the severest earthquakes possible in the area is both impractical and uneconomical. Normally, platforms are designed to withstand and minimize earthquake damage to the extent possible for moderate earthquakes that can be expected during the lifetime of the structure.

2.1.3.3.3 Ice and Snow-induced Load

Ice and snow-induced loading is a major threat for offshore structures in arctic and subarctic zones. Moving ice blocks in colder regions poses the threat of a collision with a structure. Adequate safety measures are incorporated during the design phase to mitigate both impact load caused by collision and loading caused by snow and ice formation.

2.1.3.3.4 Temperature-induced Load

Any change in temperature results in thermal stresses in conductive materials. Temperature profiling of air and sea water from the mean sea level to the seabed is carried out for mitigating such issues during the design phase. Moreover, the temperature profiling of product and product processing is also important to mitigate problems that may occur during the operation phase caused by temperature changes during the design phase.

2.1.3.3.5 Marine Growth

Different marine organisms, consisting of hard and soft growths, may accumulate on jacket and members in the splash zone as well as on the submerged portion to varying submergence depths. The quantity and quality of growth depends on environmental conditions. Normally, such growths occur within few months of installation. Marine growth increases the diameter or the exposed area of the affected section, thereby increasing the overall loading and drag force caused by waves on the structure. Drag force is the maximum in areas with hard growths. Normally, marine growth is the maximum near the splash zone/mean water level. Although certain allowance is provided for such growth during design, if the accumulation exceeds the allowed margin, marine growth is removed using different method by mechanical cleaning by water jets, blasting, brushing with the help of either diver or diver less method. Nowadays, growth-preventing coatings or wave/current-assisted devices are mounted at appropriate locations to mitigate this problem. Predominantly, barnacles, hydroids, bryozoa, hard coral, and mussels are hard marine growths found near and around the splash zone (transition point of the structure from air to water). Thickness of the growth may be up to 200 mm with coverage up to 100%. Normally, such hard growths are found in the upper reach of submergence up to around 40–50 m, however, it depends on the marine environment.

Predominantly, sponges, seaweed, and soft coral are soft marine growths. Typically, these are found on platform jacket structure, risers, caissons, J-tubes, and supporting clamps.

2.1.3.4 Additional Loads during Installation and Construction

Different operations are performed during installation and construction, including cutting, welding, machining, transportation, erection lifting, launching, and upending. Structural components experience temporary loading at different points during these operations. Care needs to be taken during the design phase to mitigate any adverse impact caused by such temporary loading. Figure 2.6 provides an illustration of the installation of an offshore structure at sea.

2.1.3.5 Unforeseen Load

Unforeseen loads cannot be forecasted, possibly occurring on account of exceptional events, including collision with floating structures, fire and/or explosion, and dropped objects. Normally, mitigation/remedial measures are incorporated at the time of design, construction, and installation, and sometimes even after commissioning.

2.2 FIXED PLATFORMS

Fixed platform is a stationary offshore structure extending upward from the mean sea level and connected to the seabed by piling, concrete column, or other suitable means. There are different types of fixed platforms:

1. Concrete gravity structure
2. Guyed tower or compliant tower
3. Jacketed platform
4. Jackup platform

2.2.1 CONCRETE GRAVITY STRUCTURE

Offshore concrete structures have been successfully used for approximately 30 years, and are mostly used in the petroleum industry as drilling, extraction, or storage units for crude oil or natural gas. These large structures house machinery and equipment needed to drill and/or extract oil and gas.

Offshore Structure and Design

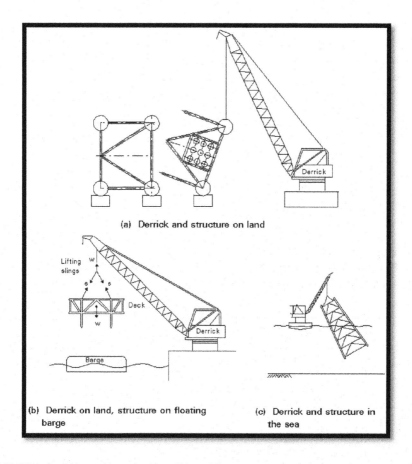

FIGURE 2.6 Lifts under various conditions [6].

Depending on the circumstances, platforms may be attached to the ocean floor, consist of an artificial island, or remain floating. In general, offshore concrete structures are classified into fixed and floating structures. Fixed structures are mostly built as concrete gravity structures (also termed as caisson-type), where the loads bear down directly on the uppermost layers as soil pressure. The caisson provides buoyancy during construction and towing and acts as a foundation during the operation phase.

Floating units may be held in position by anchored wires or chains in a spread mooring pattern. Because of the low stiffness associated with these systems, the natural frequency is low and the structure can move in all six degrees of freedom. Floating units serve as production, storage, and offloading units (FSO) for crude oil or as terminals for liquefied natural gas.

Concrete offshore platforms are almost always constructed along their vertical altitude, which allows the onshore installation of deck girders and equipment and the subsequent transport of the entire structure to the installation site [7].

A concrete gravity platform differs from template platforms in several aspects. The most fundamental difference is how the platform is anchored to the seabed. All structures, that is, template platforms tension-leg, guyed tower utilizes some form of anchoring system, that is, piles, anchored guy lines, anchored tension cables, etc. On the other hand, concrete gravity platforms sit firmly on the seabed and are held in place by the sheer force of their own weight (Figure 2.7).

Concrete gravity platforms have fixed bottom structures and remain in place on the seabed without the need for piles. Partial construction of these platforms occurs at dry docks adjacent to the sea. The structure is built from the bottom up, similar to an onshore structure. At a certain point, the dock is flooded (ballasting) and the partially built structure floats [8]. Then, it is towed to deeper sheltered water where the remaining construction is completed. After towing to the field, the base is filled with water (ballasting) to sink it on the seabed. Concrete gravity platforms are used for moderate water depths of up to 300 m [9].

Advantages and salient features:

- Requires less maintenance.
- Can be installed where the soil is firm having substantial bearing capacity and the seabed must be virtually leveled.

FIGURE 2.7 Fixed platform structure [6].

- The material of construction, that is, concrete, is relatively inexpensive and easily available.
- Has the facility for storage and requires shorter installation period as piling is not required.
- Highly fire and explosion-resistant.
- Requires less maintenance.
- Requires favorable sheltered water for construction.

2.2.1.1 Design

The large size of offshore concrete gravity structures as well as large environmental forces can cause design problems. The structural requirements for these structures include material quality, strength, and serviceability. The design is targeted to offer the least resistance to environmental factors while providing adequate support for the structure. Typically, the structure is designed to meet the criteria laid down for the ultimate progressive collapse, fatigue, and serviceability limits. Prestressed concrete, used for concrete gravity structures, provides good resistance to fatigue and corrosion. Prestressing is essential because it permits the concrete to remain compressed at all times.

The following temporary loading conditions may very well govern the structural design (Figure 2.8):

- Construction in dry dock
- Construction in protected harbor
- Ballasting for deck installation
- Towing
- Installation

Often, the cellular base walls are not pressurized, and therefore, must be designed to resist the substantial hydrostatic pressure imposed during immersion and in place condition. Coincidentally, the ballasting of concrete gravity structures for deck installation prior to towing to site is often regarded as an effective, full-scale, onshore pressure test prior to offshore installation [10].

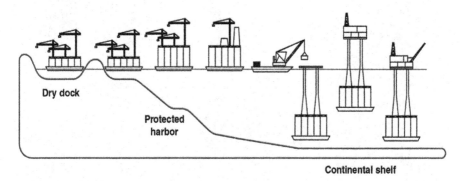

FIGURE 2.8 Concrete platform construction and installation [6]. (Courtesy of MSL engineering.)

2.2.1.2 Construction and Installation

The construction and installation of concrete gravity structures are entirely different from those of jacket structures. Figure 2.8 shows a typical set of construction and installation steps for a concrete gravity structure. As illustrated, the base of the structure is constructed in a dry dock, after which it is floated out and moored in a deepwater protected harbor [11]. Completion by slip-forming of all the cellular base walls is undertaken in the harbor, followed by slip-forming of the towers in a continuous process (Figure 2.9). Once the towers are constructed and topped off, the whole structure is ballasted down to receive the topsides deck and modules. The completed platform is de-ballasted to a minimum draft for towing and is towed using tugboats to its final location and ballasted onto the seabed. It can be observed that offshore hook-up is minimized because most of the topsides equipment and facilities are commissioned onshore prior to placement on the deck [12].

FIGURE 2.9 Construction of the troll platform [13]. (Courtesy of Photographic Services, Shell Intl., London.)

Offshore Structure and Design

Advantages:

- Concrete structures lower maintenance costs
- Long life of the structure
- Large storage capacity
- Support large deck loads

Disadvantages:

- Cost increases exponentially with depth
- Foundation settlement
- Subject to seafloor scour

2.2.2 Compliant Tower and Guyed Tower

Compliant towers reduce the steel requirements by using a jacket of constant cross-sectional area. These towers use guy lines, articulated columns, and buoyancy tanks to maintain position (Figure 2.11). Compliant towers differ from fixed platform jackets in that they are configured to respond flexibly to large waves. Compliant towers can support a large number of platform wells and may serve as host to subsea wells. Risers are supported laterally along their length and can be set at close spacing without the risk of interference. Compliant towers are mostly installed in water depths ranging from 1,500 to 3,000 ft (450 to 900 m). These towers are designed to sustain hurricane conditions. Significant lateral deflections and forces are experienced by compliant towers in case of hurricanes. Chevron Petronius tower installed at a water depth of 623 m is the deepest tower at present. Flex legs or axial tubes types of flex elements are used to reduce resonance and de-amplify wave forces. These flex elements can be designed to adapt to existing fabrication and installation equipment [14].

Compared with floating systems, such as tension-leg platforms and SPARs, the compliant tower has a sturdier design to withstand significant lateral deflections during hurricane conditions [15].

During fabrication and installation of equipments, this type of installation structures can be configured.

The freestanding compliant tower concept (Figure 2.10, adopted for the Petronius platform in the Gulf of Mexico) and the guyed tower concept (Figure 2.11, adopted for the Lena platform in the Gulf of Mexico) are based on the principle of "compliance" to attack by waves. Vertical and lateral stability is assured through the use of either flotation devices or guys. The dynamic response of a compliant tower is crucial, and dynamic analyses form a part of the design process for this structure.

2.2.2.1 Design

The surface facility of the tower comprises the drilling, production, and crew quarter modules. Individually, size is determined by the dimensions needed to handle production, drilling operations, and crew accommodations. The surface facilities are smaller on compliant towers than those on fixed platforms because of the decreased jacket dimensions.

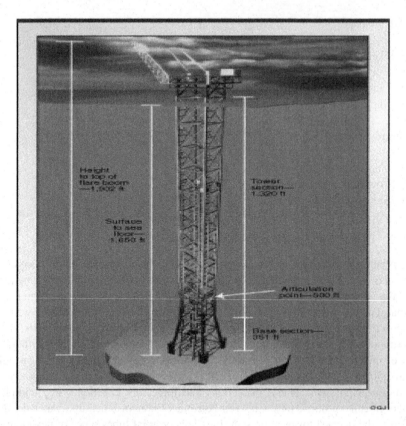

FIGURE 2.10 Compliant towers [16].

For a compliant tower, the supporting structure may have a lower and upper section. Typically, the tower's jacket is composed of four leg tubular that can range from 3 to 7 ft in diameter and are welded together with pipe braces to form a space-frame-like structure. The lower jacket is secured to the seafloor by weight and with 2- to 6-ft piles that penetrate hundreds of feet beneath the mud line. Both the lower and upper jacket dimensions can range up to 300 ft on one side. The water depth that the structure will reside in dictates the height of the jacket.

A series of buoyant tanks (up to 12) located on the upper part of the jacket places the members in tension, reducing the foundation loads of the structure. The tanks can range up to 20 ft in diameter and 120 ft in length. The amount of buoyancy is digitally programmed for control maintaining the appropriate tension in the structure members during wind and wave movements. This buoyant system can also be incorporated into some member designs, minimizing the size and placement of the tanks.

2.2.2.2 Construction and Installation

In general, for compliant towers, mooring is only used in the guyed-tower design. For guyed towers, several mooring lines (up to 20 lines measuring 5½-inch dia.) are attached to the jacket close to the waterline and are spread out evenly around it

Offshore Structure and Design

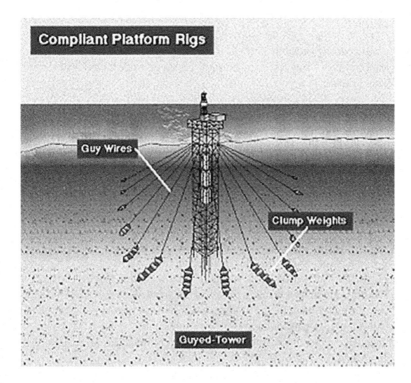

FIGURE 2.11 Guyed tower concept [17]. (Photographic archives, ExxonMobil.)

(up to 4,000 ft of line). Clump weights (120 ft × 8 ft, up to 200 tons) may be attached to each mooring line and move as the tower moves with the wind and wave forces. To better control the tower motions, the lines are kept in tension during the swaying motions. The portion of the lines past the clump weights are anchored onto the seafloor with piles (as many as 20, each 72-inch dia., 115-ft long, penetrating 130 ft, and weighing up to 60 tons). During the onshore fabrication of the jacket, the mooring system for the guyed tower is installed.

A specially designed, dynamically positioned crane barge can be utilized for installing the compliant tower, including guy line winch module, anchor pile module, and clump weight module. The installation procedure starts with the anchor piles, followed by the chains, the clump weights, and finally, the remaining mooring lines that attach to the jacket, as needed. Until the jacket is installed, anchor buoys hold the remaining lines in position [18].

Fit-for-purpose barge or ship is used to tow the jacket as per requirement in one or two pieces to the site for launching either from the side or rear of the barge/ship for installation. Typically, guyed-tower design jacket is in one piece whereas normal compliant tower has two jackets – upper and lower. Topside facilities are transported separately from the shore to the site for placement on the installed jacket.

A dynamically positioned pipe-laying ship installs the pipeline or pipeline bundle. The complete installation process can take up to 8 months before any additional drilling or production can start [16].

Advantages:

- Lower cost than steel jackets
- Good stability – guy lines and clump weights give additional restoring force
- Possible reuse

Disadvantages:

- High maintenance costs
- Small fields only
- Difficult mooring
- Cost increases exponentially with depth

2.2.3 Jacketed Platform

- Underwater, the piles are contained inside the legs of a "jacket" structure serving as bracing for the piles against lateral loads.
- The jacket also serves as a template for the initial driving of the piles. (The piles are driven through the inside of the legs of the jacket structure.)
- Natural period (usually 2.5 s) is kept below wave period (14–20 s) to avoid amplification of wave loads.
- Ninety-five percent of offshore platforms around the world are jacket-supported [19] (Figure 2.12).

FIGURE 2.12 Jacket platform [6].

Offshore Structure and Design

Typical Design: Jacket design depends on many factors including water depth, bathymetry, subsea soil condition, topside load, expected field life, and environmental conditions. Pile design, including its foundation and installation, is one of the most important factor influencing the entire jacket platform.

Installation: Offshore installation varies with the type of development involved. Fixed platforms require considerable efforts to install; jackets and topside decks need to be transported to the well location. Once the jacket is in its final position and piled, the deck is lifted into position and welded to the jacket. Following its installation, the hook-up is completed to services and production lines on the jacket. Any additional modules are lifted into position, fixed, and installed.

Topside Installations: Normally, topside facility is fabricated onshore and subsequently transported to offshore sites on a barge or ship. A topside facility can be installed in a single lift or modular lifts, and depends on the available/planned logistics and lifting capacity. Another important aspect of topside facility installation and hook-up activity is offshore weather conditions. Figure 2.13 shows the different stages in the installation of the jacketed platform, and Table 2.1 lists the activities performed in these stages.

FIGURE 2.13 Installation steps for jacketed platform [20].

(Continued)

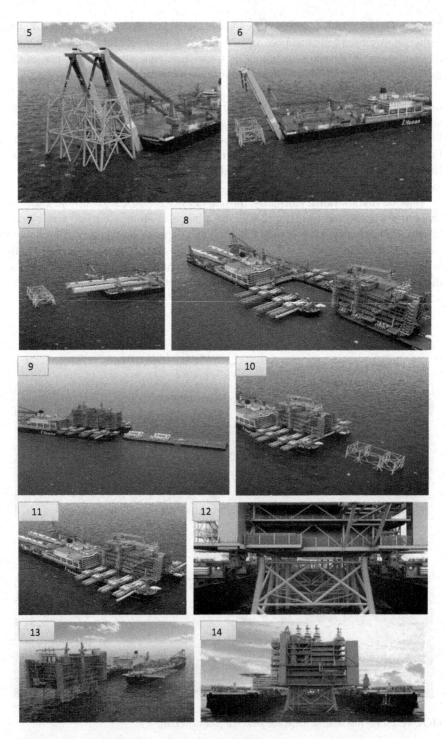

FIGURE 2.13 (CONTINUED) Installation steps for jacketed platform [20].

Offshore Structure and Design

TABLE 2.1

Stages of the Installation of the Jacketed Platform

Tag	Activity to Be Carried Out
1	The cargo barge is positioned for jacket transfer
2	The vessel sails to the installation location
3	The jacket reaches the installation location
4	The jacket is tilted to a vertical position
5	The jacket is ready to lower down
6	The jacket is lowered to the seabed
7	The jacket is positioned to the seabed, and piling and installation are started
8	The vessel approaches the cargo barge
9	The topside is loaded on the vessel
10	The vessel approaches the installation location
11	The vessel moves around the jacket
12	Motion are compensated to put topside on the jacket
13	Lowering of the topside onto the jacket
14	All support yokes are disconnected and the vessel moves away from the platform

Advantages:

- Deck load is not a constraint
- Ease of transportation and installation
- Suitable for long field life and big fields
- Processing and stabilization of the product can be done offshore
- Better stability
- Operation less dependent on water

Disadvantages:

- Cost dependent on water depth
- Prone to corrosion
- Limited logistics and storage area

2.2.4 JACKUP PLATFORM/RIG

See Figure 2.14.

2.2.4.1 Design

Once jacked up, a jackup platform/rig must remain stable under adverse environmental conditions. Design engineers should ensure that the platform's/rig's gravity load can withstand the horizontal forces of wind, wave, and ocean current. Wind provides the majority of the overturning momentum. Wind acts primarily on the portion of the legs above the hull, and the load increases as the square of wind speed. Meteorologists categorize wind speed according to the time interval used to

Jackups, as the name suggests, are platforms that can be jacked up above the sea using legs which can be lowered like jacks.

These platforms are typically used in water depths up to 400 feet, although some designs can go to 550 feet depth.

They are designed to move from place to place, and then anchor themselves by deploying the legs to the ocean bottom using a rack and pinion gear system on each leg.

FIGURE 2.14 Jackup platforms [21].

record it. The 3-second gust, defined as the highest, is used for designing individual steel members. The 1-minute wind speed is used to calculate rig stability.

Wave and ocean current are minor contributors to overturning incidents associated with jackup platforms. Wave forces depend on the height and period of the waves and water depth. Empirical formulas predict two components – inertia and drag – of the wave force against the legs assuming the worst-case scenario, the 50-year wave. Water depth influences these forces because it affects the shape of the wave.

Ocean current load is hard to predict because its velocity and direction vary with depth. Current drag can be important. Safe operation, however, depends on local conditions; if the local conditions exceed the equivalent design criteria, platform/rig stability is re-evaluated. This assessment uses laboratory data and lengthy model calculations.

Leg penetration is predicted before the rig is relocated by taking a sample of the ocean floor and studying its shear strength. Safe operation is limited by leg length. In small water, small penetration causes the leg to extend high above the jack house and increase the horizontal wind load. A minimum leg reserve of 5 ft above the top of the jack house is required as the penetration estimation is not exact.

Water depth corresponds to the highest annual tide plus a storm surge, which is a rise in the ocean level caused by low barometric pressure. This information is obtained through the local meteorological office. The air gap, measured from the mean water level to the bottom of the hull, must be sufficient to avoid the crest of any wave striking the hull. For safety, an extra 4 ft or 10% of the air gap is added [21].

2.2.4.2 Installation

The jackup platform is installed in stages illustrated in Figure 2.13. The activities as per the stage are:

A. Arriving on the location
B. Lowering leg
C. Coming out of water

Offshore Structure and Design

D. Preloading and ballasting of spud tank, and checking stability and leg penetration

E. All full air gap with environmental loads (Figure 2.15)

Advantages:

- Jackups can be easily relocated
- They have all the advantages of a fixed platform in shallow water, and do not require moorings. They have a low abandonment cost and can be returned to drilling
- Wells and riser can be of conventional type (Figure 2.16)

Disadvantages:

- Limitations on topside weight and water depth operating range
- Limited to areas where soil conditions permit satisfactory leg support
- Fatigue problems could limit utilization unless costly alterations are made to the structure
- Limited storage capability

2.3 FLOATING PLATFORM

2.3.1 MOORING AND ANCHORING

Ships or vessels at sea have to be kept in their place to enable oil or gas production, transfer, and storage on board. Mooring is a system of permanent or temporary station keeping at sea. A vessel is said to be moored when it is fastened or held secured to a fixed object such as a pier or quay or to a floating object such as an anchor buoy using cables, anchors, or lines. Mooring and anchors are very critical for the stability of a floating system (Figure 2.17).

Anchors: The mooring system relies on the strength of the anchors. The holding capacity of anchors depends on the digging depth and the soil properties. The mooring lines run from the vessel to the anchors on the seafloor. Anchor types include [23]:

- Dead weight
- Drag embedment anchor
- Pile
- Suction anchor
- Vertical load anchor

1. **Dead weight:** Dead weight is probably the oldest anchor in existence. The holding capacity is generated by the weight of the material used and partly by the friction between the dead weight and the seabed. Steel and concrete are common materials used today for the construction of dead weights.

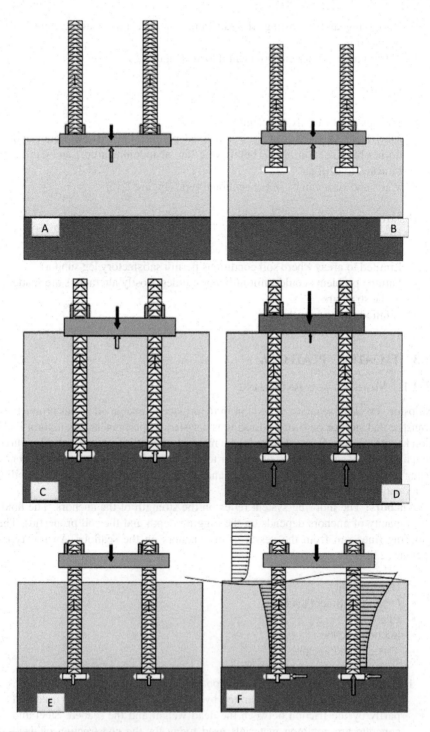

FIGURE 2.15 Jackup platform installation [22].

Offshore Structure and Design

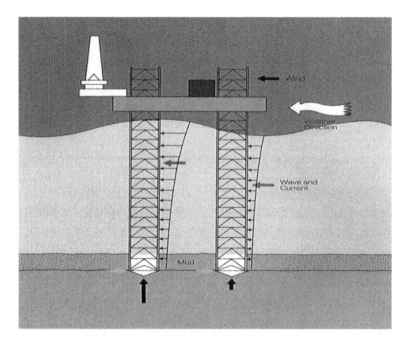

FIGURE 2.16 Various loads on the legs of the jackup platform [21].

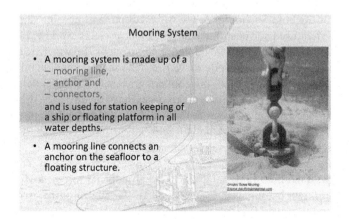

FIGURE 2.17 Mooring and anchoring [23].

2. **Drag embedment anchor:** A drag embedment anchor (DEA) is the most utilized anchor for mooring floating mobile offshore drilling units (MODUs) in the Gulf of Mexico. Drag anchors are dragged along the seabed until they reach the required depth. As they penetrate the seabed, they utilize soil resistance to hold the anchor in place. A DEA is mainly used for centenary moorings, where the mooring line arrives on the seabed horizontally, and does not perform well under vertical forces (Figure 2.18).

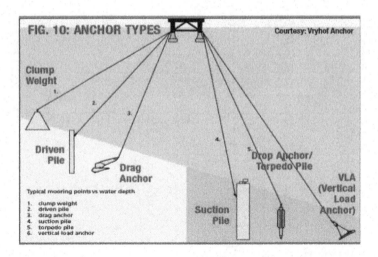

FIGURE 2.18 Different types of anchoring systems [24]. (Courtesy: Vryhol Anchor.)

3. **Pile:** The pile is a hollow steel pipe installed into the seabed using a piling hammer or a vibrator. The pile is held in place by soil friction along the pile and lateral soil resistance. In general, piles need to be installed at great depth below the seabed to achieve the required holding capacity. Piles are capable of resisting both horizontal and vertical loads.
4. **Suction anchor:** Suction piles are the predominant mooring and foundation system used for deepwater development projects worldwide. Tubular piles are driven into the seabed and a pump sucks the water from the top of the tubular, pulling the pile further into the seabed. Suction piles can be used in sand, clay, and mud soils, but not gravel, as water can flow through the ground during installation, making suction difficult. Once the pile is in position, the friction between the pile and the soil holds it in place. It can resist both vertical and horizontal forces.
5. **Vertical load anchor:** Vertical load anchors are similar to drag anchors as they are installed in the same manner. However, the vertical load anchor can withstand both horizontal and vertical mooring forces. It is used primarily in taut leg mooring systems, where the mooring line arrives at an angle to the seabed [25].

2.3.1.1 Classification of Mooring Systems
See Figure 2.19.

2.3.1.2 Catenaries Mooring System
The catenaries system is the most common type of mooring system employed in shallow water. The catenaries refer to the shape that a free hanging line assumes under gravity (Figure 2.16). The catenaries system provides restoring forces through the suspended weight of the mooring lines and its change in configuration arising from vessel motion [27]. In other words, under environmental loadings, the moored vessel

Offshore Structure and Design

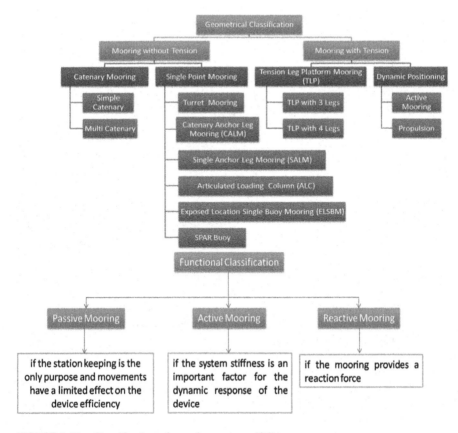

FIGURE 2.19 Classification of mooring systems [26].

tries to lift the mooring lines, which creates a restoring force mostly through the weight of the catenaries [25]. By catenaries system the mooring line terminates at the seabed horizontally, the anchor point is only subjected to horizontal forces at the seabed. This requires that the mooring lines be relatively long compared to the water depth. With the increase in water depth, the weight and the length of the mooring line start to increase rapidly. In deepwater, the weight of the mooring lines becomes excessive and the mooring lines tend to hang directly down from the rig. The excessive weight diminishes the working payload of the vessel of the floating offshore structure. To overcome this problem synthetic ropes are used in the mooring line (Figure 2.20).

2.3.1.3 Single Point Mooring System

In the single point mooring system, mooring lines are connected to a single point. Single point mooring systems are generally used on ships, as shown in Figure 2.17. The single point can consist of an external or internal turret, a floating buoy in a catenary anchor leg mooring (CALM), or a single anchor leg mooring (SALM). The single mooring system allows for a ship to weathervane into environmental conditions. The ship is often free to rotate 360°.

FIGURE 2.20 Example for mooring system (www.dredgingengineering.com) [27].

Single Point Mooring (SPM) Buoy serves as a link between the offshore facilities and the tankers for loading or offloading liquid and gas cargo. Some of the major benefits of using SPM include [28]:

- Ability to handle extra-large vessels
- Does not require ships to come to the port, saving both fuel and time
- Ships with high drafts can be moored easily
- Large quality of cargo can be easily handled

Mooring and anchoring system, buoy body, and product transfer system are the main parts of the SPM Buoy. The mooring arrangement is such that it permits the buoy to move freely within the defined limits, considering wind, waves, current, and tanker ship conditions. The SPM Buoy is anchored to the seabed using anchor chains (legs) attached to the anchor point (gravity-based or piled) on the seabed. Chain stoppers are used to connect the chains to the buoy.

Because floating platforms are normally used in deep and ultradeep waters, types of floating platforms, design, installations, and application considerations will be further discussed in detail in Chapter 6 (Deep Water Development System).

REFERENCES

1. *"How Offshore Drilling Works"*. Available at https://science.howstuffworks.com/environmental/energy/offshore-drilling.htm.
2. W.L. Leffler, R. Pattarozzi, and G. Sterling, *Deepwater Petroleum Exploration & Production*, PennWell Corporation, Tulsa, OK (2003).
3. H. Ivanov, *Corrosion Protection Systems in Offshore Structures*, Honors Research Projects. (2016).
4. *offshore-platformdesign*. Available at https://es.slideshare.net/rahulranakoti/offshore-platformdesign.

Offshore Structure and Design

5. N. Haritos, Introduction to the analysis and design of offshore structures—an overview, *Electron. J. Struct. Eng.* 7 (2007), pp. 55–65.
6. *Minimal Offshore Facilities of the Future. 2001. League City, Texas: PennWell Conferences and Exhibitions* (2001).
7. *Offshore_concrete_structure.* Available at https://en.wikipedia.org/wiki/Offshore_concrete_structure.
8. M. Waller and M.A. Shah, Advances in drilling technology, *Trans. Inst. Min & Metall. Sect. A: Min. Ind.* 101 (1992), pp. A166–172.
9. *The deepest offshore drilling rig in the world.* Available at http://www.energy industryphotos.com/deepest_offshore_oil_drilling_ri.htm
10. *Type of offshore structures, Retrieved from.* Available at http://omeleteholic.blogspot. com/search?q=type-of-offshore-structures
11. C. Günther, E. Lehmann, and C. Österfaard, *Offshore Structures: Conceptual Design and Hydromechanics*, Springer, Berlin (1992).
12. C. Hard, Spoilt for choice: How to classify and select minimum facility solutions, in *Conference on Minimal Offshore Facilities of the Future* (2001).
13. P. O'Connor, S. Defranco, and B. Manley, *Minimal Structures Open Global Production Opportunities*, Offshore Magazine (1999).
14. S. Chakrabarti and C. Capanoglu, Learn more about compliant tower historical development of offshore structures, in S. Chakrabarti (ed.) *Handbook of Offshore Engineering*, Offshore Structure Analysis, Inc., Plainfield, IL (2005).
15. *Concrete platform construction and installation.* Available at http://blog.gooshared. com/view/166.
16. *Compliant Tower: The Next Generation.* Available at https://www.offshore-mag. com/business-briefs/equipment-engineering/article/16757589/compliant-towers-the-next-generation.
17. G. Moritis, Tallest structure installed in gulf, *Oil Gas J.* 96 (1998), pp. 49–52.
18. *platform-compliant-tower.* Available at https://www.globalsecurity.org/military/systems/ship/platform-compliant-tower.htm
19. *31-story-about-platform.* Available at http://bourneronin.blogspot.com/2010/06/31-story-about-platform.html
20. *offshore/steel.* Available at www.esru.strath.ac.uk/EandE/Web_sites/98-9/offshore/steel. htm
21. *BENNETT & ASSOCIATES.* Available at www.keppelfels.com.sg (A technical primer by BENNETT & ASSOCIATES, L.L.C.).
22. *How Do Jackups Work?* Available at www.rigzone.com/training/insight.asp?insight_id=339&c_id=
23. *How do Mooring Systems Work?* Available at www.rigzone.com/training/insight. asp?insight_id=358
24. *Anchor manual 2005, Vryhof anchors.* Available at www.dredgingengineering.com/dredging/media/LectureNotes/Anchors/AM2000.pdf
25. *No Title.* Available at http://svrmoorings.weebly.com/services-we-provide.html.
26. AEA Technology Engineering Solutions, *Failure modes, reliability and integrity of floating storage unit (FPSO, FSU) turret and swivel systems*, Offshore Technol. Rep. 2001/073 (2001).
27. B. Zanuttigh, L. Martinelli, and M. Castagnetti, *Screening of Suitable Mooring Systems*, Aalborg University, Aalborg (2011), pp. 1–31.
28. *Offshore Mooring Lines, by Offshore Consulting Engineering.* Available at www. dredgingengineering.com/moorings/lines/Offshore mooring lines mooring system. htm.

3 Offshore Drilling and Completion

Drilling is undertaken once the necessary support system (drilling platform) along with living accommodation for the drilling crew is in place. This chapter focuses on drilling and completion operations offshore.

3.1 OFFSHORE DRILLING

3.1.1 INTRODUCTION

Drilling an offshore well is much more challenging compared to drilling an onshore well. At the same time, economics associated with offshore drilling operations are quite significant given their complexity and difficulty. Because the platform cost is significantly high, separate platforms for individual well and drilling a vertical well may be cost prohibitive. Directional drilling has made it possible to drill a large number of wells in each direction with only a few vertical wells, significantly reducing the cost of field development. Rigs used for offshore drilling are classified into two basic types and are discussed in Chapter 2.

- Bottom supported
  Fixed – platform
  Mobile (MODU) – Jackup, barge, submersible
- Floating
  Fixed – Tension-leg platform, SPAR
  Mobile (MODU) – Semi-submersible, drillship [1].

Offshore rigs and land rigs have several similarities in terms of type of equipment used. While drilling depth dictates the equipment design in land rig; it is the water depth that largely dominates in deciding the type of platform to be used for installation of rig in offshore. Matching drilling units to platform type/vessels for different water depth is depicted in Figure 3.1.

Jackup rigs are the most widely used drilling rigs in offshore drilling operations. A jackup rig is a barge fitted with long support legs that can be raised or lowered. The jackup is maneuvered (self-propelled or towed) into location with its legs up and the hull floating on the water. Upon arrival at the desired location, the legs are jacked down onto the seafloor. Subsequently, "preloading" takes place, where the weight of the barge and additional ballast water are used to drive the legs securely into the sea bottom such that they do not further penetrate during operations. After preloading, the jacking system is used to raise the entire barge above water to a predetermined

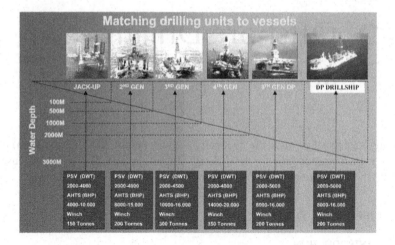

FIGURE 3.1 Matching drilling units to vessels (Courtesy: Greatship India Limited).

height or "air gap", so that wave, tidal, and current loading acts only on the relatively slender legs and not on the barge hull [2].

Because offshore drilling operations involve drilling multiple wells from a single platform, use of directional well drilling and extended reach drilling (ERD wells) is increasing to penetrate long intervals of pay zone and achieve more hydrocarbon exploitation from a single platform. As the difficulty of drilling such advanced wells is high compared to conventional well drilling, special completion type is required for such wells. Types of completion used in offshore operations is discussed in Section 3.2.

Because offshore operations are very expensive, the first and most important step is efficient planning.

3.1.2 Well Planning [3]

Well planning refers to the identification of all the required resources for drilling, completion, and production, as well as utilizing these resources effectively and efficiently for production throughout the life cycle of the well.

Well planning is a multidisciplinary exercise consisting of mud program, casing program, drill string design, bit program, etc., and the best approach in well planning is the one in which service contractors become equally involved with their area of expertise [4].

Well planning process includes:

- Setting priorities
- Information flow
- Decision support
- Optimization

Wellbore planning requires consideration of various parameters such as completion, trajectory, well stability, drilling fluid, casing, cementing, wellhead/BOP, BHA and

Offshore Drilling and Completion

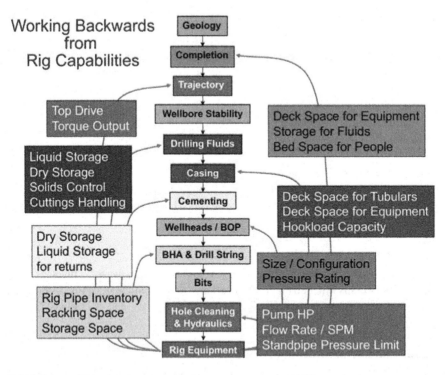

FIGURE 3.2 Well planning process. (Courtesy: www.petroskills.com.)

drill string, drill bits, hole cleaning, and hydraulics based on geology, reservoir, and suitable rig equipment (Figure 3.2).

3.1.2.1 Geology
Geology of formation must be studied to determine rock type/strength, porosity/permeability, dip angles/faults, fracture and pore pressure gradients, expected fluid, and expected flow rates. This can be achieved with the help of offset well data, electric logs, mud logs, samples, cores, test results, and seismic data. With the help of geology, target selection and specifications can be obtained, including true vertical depth (TVD), size, shape, orientation, directional tendencies, dip angles, and faults.

3.1.2.2 Completion Design
Parameters to be considered for completion design include production casing/tubing size, completion design, artificial lift requirements, stimulation design, length/orientation in reservoir, and future intervention requirements.

3.1.2.3 Trajectory Design
Trajectory design include measured depth (MD)/TVD, inclination, Azimuth, kick of point (KOP), build rates, walk rates, dogleg Severity, N-S/E-W displacements, horizontal departure, anti-collision requirements, and surveying requirements.

3.1.2.4 Wellbore Stability

Well is designed such that wellbore stability is maintained. Different stability parameters include mechanical stability, mud weight (MW) window, casing seat options, drilling practices, tripping chemical stability, and inhibitive fluid requirements

3.1.2.5 Drilling Fluid Design

This includes mud type, mud properties, mud weight, rheology, filtration, inhibition, solids control requirements, and handling/disposal of cuttings.

3.1.2.6 Casing Design

This includes casing sizes, weights, grades, connections, casing setting depths, casing versus liner/tieback, and load cases – tension, burst, collapse, buckling, special tools, and running procedures.

3.1.2.7 Cement Job Designs

This includes slurry type/additives, slurry volumes, spacers/washes, displacement, conventional versus inner string, tool selection – float equipment, centralizers, stage tools, external casing packers (ECPs), formation integrity test (FIT)/leak off test (LOT) plan.

3.1.2.8 Wellhead/Blowout Prevention (BOP) Design

This includes wellhead size/design/pressure rating, blowout prevention (BOP) size/configuration/working pressure rating, and testing procedures/frequency.

3.1.2.9 BHA and Drill String

These include bottom hole assembly (BHA) configuration/directional response; drill string – length, size, weight, grade, connection; performance properties – tension, torque, buckling; operating procedures – torque and drag, percent rotary mode, tortuosity, casing wear prediction/mitigation plan, inspection procedures/frequency.

3.1.2.10 Bit Design

This includes bit size, roller cone/fixed cutter, style, features; operating parameters – WOB, RPM, expected ROP, footage, bit life, cost/foot, eccentric or bicenter requirements.

3.1.2.11 Hole Cleaning and Hydraulics

These include flow rates, standpipe pressures, pump horsepower requirements, annular velocities, equivalent circulating densities, surge/swab, bit hydraulic horsepower, hole cleaning recommendations.

3.1.2.12 Rig Equipment

This includes top drive torque/RPM capability, pump HP/flow rate/pressure capability, hook load capacity, Derrick setback space/capacity, drilling fluid storage/mixing/solids control, bulk dry/liquid storage capacity, deck space/bed space.

Offshore Drilling and Completion

3.1.3 RIG SELECTION CRITERIA [3]

Rig selection criteria include the following:

- Health, safety, and environment (HSE) compatibility
- Technical capability
- Full-cycle efficiency
- Cost
- Availability

3.1.3.1 HSE Compatibility

Lost time injury (LTI) rate; deck space to minimize crowding; variable load to minimize crane and boat operations; zero discharge – cuttings drying/storage/offloading, slurrification, and reinjection system; and familiarity.

3.1.3.2 Technical Capability

This includes access to all platform slots from primary access; load limits at cantilever distance;

pumps – number, HP, pressure, drill pipe, and other tubulars; inventory; and handling

systems.

3.1.3.3 Full-Cycle Efficiency

This includes transitions – drilling to completion, handling multiple fluids/tubular strings, minimum stack changes for required pressure, rig move, sea states and procedures, rig up/rig prep procedures, accommodations, and historical non-productive time.

3.1.4 WELLBORE STABILITY

The successful completion of gas and oil wells involves the selection of mud weight to maintain hole stability, avoid formation fluid intrusion into the wellbore, and minimize mud loss to the formation. Our concern here is to understand the rock mechanical response to drilling and hydrostatic fluid support. Borehole instability is caused by the tensile or compressive failure of the borehole wall. The borehole fails in tension when the pressure exerted by the drilling mud induces stresses in the borehole wall that exceeds the tensile strength of the rock. The failure manifests in the form of cracks, typically starting from the borehole and running radials inside formation [5]. Drilling mud may then penetrate and propagate through these cracks, leading to a fall in mud level in the borehole. If this continues, the borehole stability is eventually restored by the resulting reduction in the hydrostatic load of the hole at the depth. The borehole fails in compression when the pressure of the drilling mud is insufficient to maintain the shear stresses in the borehole wall below the shear strength of the formation [5]. When the borehole fails in compression, broken rock falls into the borehole, and the borehole diameter increases at the point of failure. Both the increase in borehole diameter and the volume of rock debris falling into the borehole sometimes make it difficult or impossible to move drilling equipment into or out of the borehole. Certain rock types, such as salt, creep rather than fail when compressed and may close around equipment in the borehole or reduce borehole diameter, again making it difficult or impossible to move drilling equipment into or out of the hole. The implications of borehole instability to lost drilling time and equipment has prompted operators and service companies to apply rock mechanics principles to define working limits for mud weights to avoid tensile or compressive failure [5]. This is particularly true for long reach, highly deviated, and horizontal wells where the cost of downtime is very high. The theoretical analysis involved in borehole stability is quite complex and requires a great deal of mathematical derivation.

The quantification of wellbore instability requires understanding and quantifying the following five steps:

1. Determining the magnitude and direction of in-situ earth stresses
2. Determining rock properties
3. Establishing a rock failure criterion
4. Calculating induced stresses around the wellbore for vertical and deviated wells
5. Comparing induced stresses with the stresses from failure criterion to establish if the wellbore will fail [6].

3.1.4.1 Stress Distribution around the Wellbore

The in-situ stresses in an undisturbed rock mass are shown in Figure 3.3. It is usually assumed that the major principal stress, σ_1, is vertical and the intermediate and minor principal stresses, σ_2 and σ_3, respectively, are horizontal. However, in tectonically active areas, the principal stresses may be inclined vertically and horizontally. When a borehole is drilled into a rock mass, drilling fluid of a different density replaces the excavated rock, and the natural stress field redistributes locally around the borehole. The stress distribution in the borehole wall depends on the magnitude

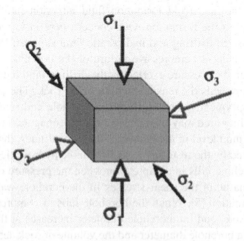

FIGURE 3.3 In-situ stresses in an undisturbed rock [7].

Offshore Drilling and Completion

of the in-situ principal stresses and the stress–strain response of the rock material. Borehole stability studies have assumed a homogeneous, isotropic, linear elastic rock, with no fluid flow through. Hence, drilling a hole in the ground disturbs the in-situ stresses around the wellbore and induces additional stresses. In particular, hoop or circumferential stresses, $\sigma_{\theta\theta}$, are induced which act around the wellbore. The mud pressure creates radial stresses, σ_{rr}, which provide support for the walls of the wellbore. As σ_{rr} increases, the induced hoop stresses decrease and may become negative, resulting in rock failure in tension, that is, wellbore burst. A third stress, σ_{zz}, longitudinal stress, acts along the axis of the wellbore. Hence, at any point near the wellbore, there will be three induced stresses – $\sigma_{\theta\theta}$, σ_{rr}, and σ_{zz}. These stresses are mutually perpendicular to each other at any point [8].

3.1.4.2 Establishing a Minimum Safe Mud Weight [9]

Which criterion is used to establish the minimum safe mud weight? Clearly, it is one that will minimize the risk of complete hole collapse [10]. However, additional factors can also influence this, including:

- The volume of cuttings
- The inclination of the well
- The position around the well of the breakouts

The cuttings volume and well inclination are important because of hole-cleaning issues. The larger the cuttings volume per unit hole length, the better the hole cleaning needs to be. Because hole cleaning is easier in vertical wells than in deviated wells, vertical wells can accommodate larger amounts of failure.

Increase in pumping rate and carrying capacity, or reduced penetration rates, can mitigate the risk associated with excessive cuttings volumes. Because in deviated wells there is considerable pipe contact with the top and bottom of the well, breakouts in these locations are likely to be more problematic than breakouts on the sides of the hole. However, if the well needs to be steered, breakouts on its sides may adversely affect directional control. Because breakout width is relatively easy to measure and is directly related to cuttings volume, and because breakout depth increases with time, we ordinarily choose the breakout width as the criterion to establish the appropriate minimum mud weight [11].

Because breakouts have been observed that extend more than 100° on each side of a well in vertical wells drilled into some shales, this is an appropriate limit for such wells. Narrower breakouts will become problematic in more brittle rock, therefore, in practice, it is best to use a breakout width limit of 90° for breakouts on each side of a vertical well. This limit implies that at least half of the wellbore circumference must be intact, a condition that has been referred to as "sufficient to maintain arch support" in sanding analyses. Because hole cleaning is more difficult in deviated wells, the maximum safe breakout width should be reduced as deviation increases. It is important to remember that it is not necessary to completely avoid breakout formation to drill wells safely. Using such an overly restrictive criterion is not only unnecessary but will inevitably lead to recommendations for excessively high mud weights in situations where these are not warranted.

3.1.4.3 Validating the Geomechanical Model [9]

When using geomechanical analysis, it is important to use prior drilling experience to validate the geomechanical model. This is possible, even when no log data are available for previous wells, by using drilling events such as:

- Mud losses
- Tight spots
- Places necessitating repeated reaming
- Evidence of excessive or unusually large cuttings

If wellbore stability predictions for existing wells are capable of reproducing previous drilling experience, we can be confident that the geomechanical model is appropriate for use in predicting the stability of planned wells.

Figure 3.4 was prepared using the drilling mud program for that well, and the geomechanical model was developed for the field based on offset experience.

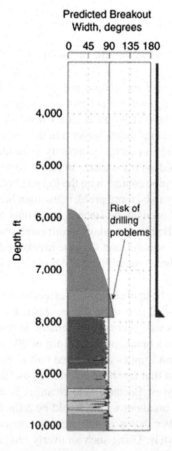

FIGURE 3.4 An example prediction of the degree of wellbore instability in a vertical well in deep water [12].

Offshore Drilling and Completion

The model indicates that, while the section above 5,800 ft will be quite stable (no failure is predicted), below that depth, failure will progressively worsen, until at 7,400 ft, it is severe enough to cause considerable drilling problems. Although the model could not explain problems encountered in this well above 5,400 ft, it turned out that these problems were not caused by geomechanics because they were mitigated with no change in mud weight, and no evidence of enlargement was found in log data from this interval. In contrast, considerable drilling difficulties were encountered just above 7,800 ft in this well that were detailed in drilling reports, including several pack-off and lost-circulation events. These problems required setting casing prematurely at that depth. Single-arm caliper logs subsequently revealed that this section was severely enlarged [13].

Below the casing point, the mud weight increased, which reduced hole instability problems in the remaining sections of the well, as predicted by the calculations. Nevertheless, there was some evidence for wellbore enlargements in caliper data in the interval below 9,200 ft, even for the higher mud weights used. These sections were those where the predicted breakout width exceeded 90°, lending support to the validity of the geomechanical model. Subsequently, the model was used to design a number of wells, all of which reached total depth (TD) without any incident

3.1.4.4 Sand Production

The two main mechanisms responsible for sand production are shear failure and tensile failure. Shear failure is usually initiated by reducing pore pressure below a critical value. Tensile failure is usually initiated by a production rate above a critical value. Sand production caused by tensile failure at high production rates is not as prolific as that caused by shear failure. Increasing the effective stress increases the likelihood of shear failure of the formation at the borehole wall but reduces the likelihood of tensile failure. Accurate knowledge of the size and directions of the in-situ principal stresses aids more accurate calculation of the effective stresses, and therefore, better predictions of the onset of sand production.

3.1.5 CASING DESIGN

3.1.5.1 Casing Program [14]

In offshore oil industry, five types of casing are used, which are discussed below.

3.1.5.1.1 Structural Casing

Structural casing is the outer string of large diameter, heavy-wall pipe installed in wells drilled from floating installations to isolate very shallow sediments from subsequent drilling and to resist the bending moments imposed by the marine riser, as well as to support the wellhead installed on the conductor casing. Structural casing refers to a short string of large diameter pipe set by driving, jetting, or drilling to support unconsolidated shallow sediments, provide hole stability for initial drilling operations, and provide anchorage for a diverter system.

3.1.5.1.2 Conductor Casing

Conductor casing is the first string set below the structural casing (i.e., drive pipe or marine conductor run to protect loose near-surface formations and to enable circulation of drilling fluid). The conductor isolates unconsolidated formations and water sands and protects against shallow gas. This is usually the string onto which the casing head is installed. A diverter or a BOP stack may be installed onto this string. When cemented, it is typically cemented to the surface or to the mud line in offshore wells.

3.1.5.1.3 Surface Casing

Surface casing is set to provide blowout protection, isolate water sands, and prevent lost circulation. It also often provides adequate shoe strength to drill into high-pressure transition zones. In deviated wells, the surface casing may cover the build section to prevent keyseating of the formation during deeper drilling. This string is typically cemented to the surface or to the mudline in offshore wells.

3.1.5.1.4 Intermediate Casing

Intermediate casing is set to isolate:

- Unstable hole sections
- Lost-circulation zones
- Low-pressure zones
- Production zones

It is often set in the transition zone from normal to abnormal pressure. The casing cement top must isolate any hydrocarbon zones. Some wells require multiple intermediate strings. Some intermediate strings may also be production strings if a liner is run beneath them.

3.1.5.1.5 Production Casing

Production casing is used to isolate production zones and contain formation pressures in the event of a tubing leak. It may also be exposed to:

- Injection pressures from fracture jobs
- Down casing, gas lift
- The injection of inhibitor oil

Must allow for enough room below the target zone to install completion [17].

3.1.5.2 Loads Encountered While Designing Casing

There are three basic forces the casing is subjected to – collapse, burst, and tension. These are the actual forces that exist in the wellbore. They must first be calculated and then maintained below the casing strength properties. In other words, the collapse pressure must be less than the collapse strength of the casing and so on. Casing should initially be designed for collapse, burst, and tension. Refinements to the selected grades and weights should only be attempted after the initial selection is made.

For directional wells, a correct well profile is required to determine the TVD. All wellbore pressures and tensile forces should be calculated using TVD only. The casing lengths are first calculated as if the well is a vertical well, and then these are corrected for the appropriate hole angle.

1. Burst:

    In oil well casings, burst occurs when the effective internal pressure inside the casing (internal pressure minus external pressure) exceeds the casing burst strength.

    At the top of the hole, the external pressure is zero, and the internal pressure must be supported entirely by the casing body. Therefore, burst pressure is the highest at the top and the lowest at the casing shoe where internal pressures are resisted by the external pressure originating from fluids outside the casing. As will be shown later, in production casing, the burst pressure at shoe can be higher than the burst pressure at surface in situations where the production tubing leaks gas into the casing.

    Burst calculations for individual casing strings
    Conductor Casing: There is no burst design for conductors.
    Surface and Intermediate Casings

    For gas to surface (unlimited kick size), calculate burst pressures as follows: Calculate the internal pressure (Pi) using the maximum formation pressure at the next hole TD, assuming the hole is full of gas (see the below Figure 3.5).

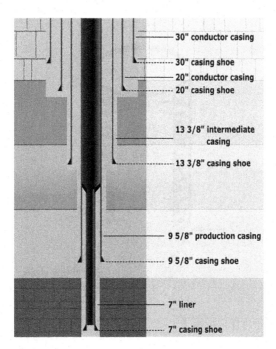

FIGURE 3.5 Types of casing [15].

Burst at surface = Internal pressure (Pi) − External pressure

Burst pressure at surface $(B1) = Pf - G \times TD$ (note external pressure at the surface is zero)

Burst pressure at casing shoe $(B2)$ = Internal pressure − Backup load = Pi − 0.465 × CSD

$B2 = Pf - G \times (TD - CSD) - 0.465 \times CSD$

The backup load is assumed to be provided by mud which has deteriorated to salt-saturated water with a gradient of 0.465 psi/ft

Production Casing: The worst case occurs when gas leaks from the top of the production tubing to the casing. The gas pressure will be transmitted through the packer fluid from the surface to the casing shoe (Figures 3.6 and 3.7).

Burst pressure = Internal pressure − External pressure

Burst at surface $(B1) = Pf - G \times CSD$

Burst at shoe $(B2) = B1 + 0.052 \, \rho_p \times CSD - CSD \times 0.465$

Where, G = Gradient of gas, usually 0.1 psi/ft

Pf = Formation pressure at production casing seat, psi

ρ_p = Density of completion (or packer) fluid, ppg

0.465 = Density of backup fluid outside the casing to represent the worst case, psi/ft.

2. Tension Criteria:

Most axial tension arises from the weight of the casing itself. Other tension loadings can arise due to bending, drag, shock loading, and during pressure testing of casing. In casing design, the uppermost joint of the string

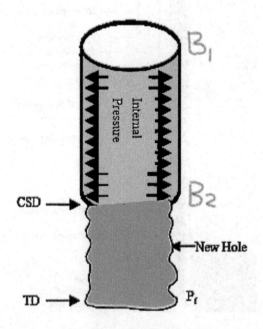

FIGURE 3.6 Burst design for surface and intermediate casing [6].

Offshore Drilling and Completion

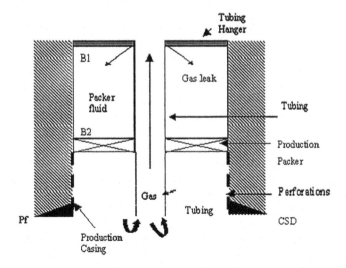

FIGURE 3.7 Burst design for production casing [6].

is considered the weakest in tension because it has to carry the total weight of the casing string. Selection is based on a design factor of 1.6–1.8 for the top joint.

Tensile forces are determined as follows:
1. Calculate weight of casing in air (positive value) using TVD;
2. Calculate buoyancy force (negative value);
3. Calculate bending force in deviated wells (positive value);
4. Calculate drag force in deviated wells (this force is only applicable if casing is pulled out of the hole);
5. Calculate shock loads due to arresting casing in slips; and
6. Calculate pressure testing forces.

Forces (1) to (3) always exist whether the pipe is static or in motion. Forces (4) and (5) exist only when the pipe is in motion. The total surface tensile load (sometimes referred to as installation load) must be determined accurately, and must always be less than the yield strength of the top joint of the casing. Moreover, the installation load must be less than the rated derrick load capacity so that the casing can be run in or pulled out of hole without causing damage to the derrick.

LOAD CASES:
There are three load cases for which the total tensile force should be calculated – running conditions, pressure testing, and static conditions. These load cases are sometimes described as installation load cases.

Load Case 1: Running Conditions
This applies to the case when the casing is run in hole and prior to pumping cement.

Total tensile force = buoyant weight + shock load + bending force

Load Case 2: Pressure Testing Conditions

This condition applies when the casing is run to TD, the cement is displaced behind the casing, and mud is used to apply pressure on the top plug. This is usually the best time to test the casing while the cement is still wet. In the past, some operators tested casing after the cement was set. This practice created microchannels between the casing and the cement and allowed pressure communication between various zones through these open channels.

$$\text{Total tensile force} = \text{buoyant weight} + \text{pressure testing force} + \text{bending force}$$

Load Case 3: Static Conditions

This condition applies when the casing is in the ground, cemented, and the wellhead installed. The casing is now effectively a pressure vessel fixed at the top and bottom. One can argue that other forces should be considered for this case such as production forces, injection forces, temperature-induced forces, etc.

$$\text{Total tensile force} = \text{buoyant weight} + \text{bending force} + \left(\text{miscellaneous forces}\right)$$

3. Collapse criteria:

Collapse pressure originates from the column of mud used to drill the hole, and acts on the outside of the casing. Because the hydrostatic pressure of a column of mud increases with depth, collapse pressure is the highest at the bottom and zero at the top.

$$\text{Collapse pressure} = \text{External pressure} - \text{Internal pressure}$$

COLLAPSE CALCULATIONS FOR INDIVIDUAL CASING STRINGS
CONDUCTOR:

The conductor is usually set at a shallow depth ranging from 100 ft to 1,500 ft. Assume complete evacuation so that the internal pressure inside the casing is zero. The external pressure is caused by the mud in which the casing was run. **For offshore operations**, the external pressure is made up of two components:

Collapse pressure at mud line = External pressure due to a column of sea water from sea level to mud line = (0.45 psi/ft) × mud line depth = $C1$ psi

Collapse pressure at casing seat = $C1 + 0.052 \times \rho_m \times \text{CSD}$

INTERMEDIATE CASING

Three collapse points will have to be calculated using the general form (Figure 3.8):

$$\text{Collapse pressure}, C = \text{external pressure} - \text{internal pressure}$$

Offshore Drilling and Completion

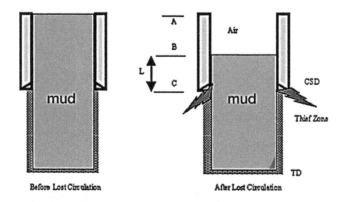

FIGURE 3.8 Collapse design for intermediate casing [6].

1. Point A: At Surface

$$C1 = \text{Zero}$$

2. Point B: At depth (CSD-L)

$$= 0.052(\text{CSD} - \text{L}) \times \rho_m - 0$$

$$C2 = 0.052(\text{CSD} - \text{L})\rho_m$$

3. Point C: At depth CSD
   $C3 = 0.052\ \text{CSD} \times \rho_m - 0.052\ \text{L} \times \rho_{m1}$ (where ρ_{m1} the mud weight in which casing was run in.)

PRODUCTION CASING

For production casing, the assumption of complete evacuation is justified in the following situations:
1. If perforations are likely to be plugged during production as in gas wells. In this case, surface pressure may be bled to zero, and hence, give little pressure support inside the casing.
2. In artificial lift operations. In such operations, gas is injected from the surface to reduce the hydrostatic column of liquid against the formation to help production. If the well pressure were bled to zero at surface, a situation of complete evacuation could exist.
3. In air/gas drilling, all casing strings should be designed for complete evacuation.
4. Another situation that results in complete evacuation is a blowout which unloads the entire hole.

3.1.6 Trajectory Design

As planning well trajectory is not a simple task at all, especially in multi-well platform, directional companies are closely involved in the planning. Aspects to be considered before calculating final well path include the following (Figure 3.9).

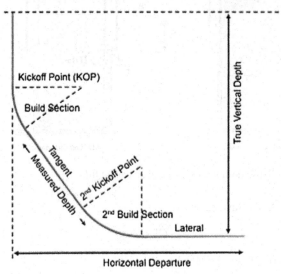

FIGURE 3.9 Trajectory components [14].

3.1.6.1 The Target

The target is usually specified by the geologist, who will define certain points as the target with acceptable tolerance (e.g., a circle of radius 100 ft having the exact target as its center). Large target zones are preferable. In the case of multiple zones penetration, they should be selected considering that planned pattern can be achieved without causing big drilling problems.

3.1.6.2 Kick-Off Point and Build-Up Rate

Measured depth from where a well is steered (usually beginning of the build section) is called kick-off point. Inclination angle difference between two consecutive survey points, extrapolated to 30 m (or 100 ft). The selection of both the kick-off point and the build-up rate depends on the following factors:

- Hole pattern
- Casing program
- Mud program
- Horizontal displacement and maximum tolerable inclination.

Kick-off points should be selected such that the well path is at a safe distance from the surrounding existing wells. The shallower the KOP and the higher the build-up rate used, the lower the maximum inclination.

Build-up rates are usually in the range of 1.5°/100' M.D. to 4.0°/100' M.D. for normal directional wells. Build-up rates selection should always be done considering maximum permissible dogleg severity.

Offshore Drilling and Completion

In practice, well trajectory is calculated for several KOPs and build-up rates, and the results are compared. The optimum choice is one which gives:

- Safe clearance from all existing wells
- Keeps the maximum inclination within desired limits
- Avoids unnecessarily high dogleg severities

3.1.6.3 Tangent Section

Tangent section is the section of a well where the well path is maintained at a certain inclination, with the intent of advancing in both TVD and vertical section. If wells are drilled at inclinations (up to 80°), the area which can be covered from a single platform is approximately eight times that covered when maximum inclination of the wells is limited to 60°. However, high inclination angles can result in excessive torque and drag on the drill string, presenting hole cleaning, logging, casing, cementing, and production problems, which are avoidable by the current technology.

Experience over the years has shown that directional control problems are aggravated when tangent inclinations are less than 15°. This is because there is more tendency for the bit to walk (i.e., change in azimuth), and so more time is spent keeping the well on course. As such, most run-of-the-mill directional wells are still planned with inclinations in the range 15°–60°.

3.1.6.4 Drop-Off Section

On S-type wells, the rate of drop-off is selected to ease casing problems and avoidance of completion and production problems. It is much less critical to drilling because there is less tension in the drill pipe that is run through deeper doglegs and less time spent rotating below the dogleg.

3.1.6.5 Trajectory Measurements

3.1.6.5.1 Inclination (or) Drift

The angle of the wellbore defined by a tangent line at any point of wellbore and a vertical line is called the inclination. The angle measured in degrees between the actual well path at some depth and a vertical line below the rig site. By oilfield convention, 0° is vertical and 90° is horizontal (Figure 3.10).

3.1.6.5.2 Azimuth

The azimuth of a wellbore at any point is defined as the direction of the wellbore on a horizontal plane measured clockwise from a north reference. Azimuths are usually expressed in angles from 0° to 360°, measured from zero north.

$$TRUE\ NORTH = MAGNETIC\ NORTH \pm (DECLINATION)$$

All magnetic tools give readings to magnetic north; however, the final calculated coordinates are referenced to either true north or grid north (Figure 3.11).

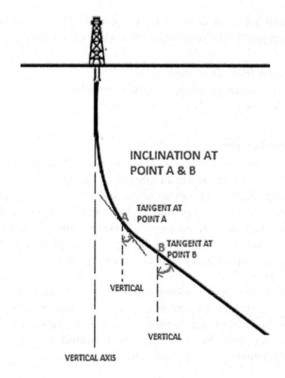

FIGURE 3.10 Inclination angle [14].

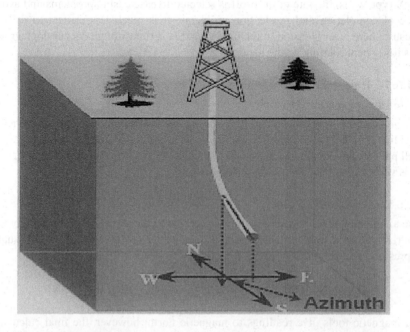

FIGURE 3.11 Azimuth [14].

3.1.6.5.3 Types of Directional Patterns [14]

The advent of steerable systems has resulted in wells that are planned and drilled with complex paths involving three-dimensional turns. It is used especially in case of re-drills and side-tracked wells. The well path should be kept as simple as possible. Different types of vertical projections for common directional patterns are shown in Figure 3.12 and Table 3.1.

3.1.6.5.4 Well Trajectory Calculation [16]
- **Build section**
  - Build radius = R_1
  - Build-up rate (BUR)
  - Length of arc = $L_1 = r_1 \cdot \Theta$
  - Vertical depth = $r_1 \cdot \sin\Theta$
  - Horizontal deviation = $r_1 \cdot (1-\cos\Theta)$
- **Hold section**
  - Length of hold section = L_2
  - Vertical depth = $L_2 \cdot \cos\Theta$
  - Horizontal deviation = $L_2 \cdot \sin\Theta$
- **Drop section**
  - Drop radius = R_2
  - Drop rate (DR)
  - Length of arc = $L_3 = r_2 \cdot \Theta_2$
  - Vertical depth = $r_2 \cdot \sin\Theta_2$
  - Horizontal deviation = $r_2 \cdot (1-\cos\Theta_2)$

Total measured depth = Kick-off depth + Length of arc of build + Length of hold section + Length of arc of drop (Figure 3.13).

Total measured depth = $D_1 + L_1 + L_2 + L_3$

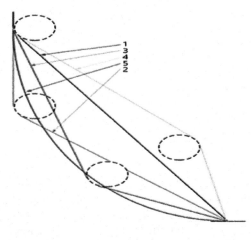

FIGURE 3.12 Types of directional well trajectory [14].

TABLE 3.1
Directional Trajectories Options, Features, Advantages, and Disadvantages

Sr. No	Trajectory Options	Features	Advantages	Disadvantages
1	Build and Hold	Constant BUR till tangent angle, and keeping the tangent angle constant	Simple, low tangent angle, large horizontal displacements can be achieved	High contact force in build can result in high torque and casing wear
2	Undersection	Build and hold with deep KOP	Lower contact force in build section that results in application in salt dome drilling, fault drilling, and side-tracking or repositioning of target	High tangent angle results in less horizontal displacement; BUR is difficult to control; more tripping time to change BHA while deflecting
3	Double Build	Build-hold-build-hold trajectory; can use two different BRs in curves.	Long horizontal displacements with less contact force in upper build section	Deep steering is required; high angle in second tangent
4	S-Shaped (or) Modified S-Shaped	Build-hold-drop (or) Build-hold-drop-hold trajectory to reach the target	Low angle reservoir entry; drilling of relief well where it is necessary to run parallel to wild well; completing a well that intersect multiple producing zones	More torque and drag can be expected due to the additional bend; Reduces final angle in reservoir bringing risk of key seating
5	Catenary	Continuously increasing BUR with depth without tangent section	Lowest contact force of any trajectory; smooth drilled wellbore, thus reducing torque and drag, less chance of key seating, and differential sticking	Not cost-effective in implementation; limited reach; hard to select BHAs with the required rate of build

- Total measured depth $= D_1 + r_1 \cdot \Theta + L_2 + r_2 \cdot \Theta_2$

  Total vertical Depth $= D_1 +$ Vertical depth of build + Vertical depth of hold section + Vertical depth of drop
- Total vertical depth $= D_1 + r_1 \cdot \sin\Theta + L_2 \cdot \cos\Theta + r_2 \cdot \sin\Theta_2$

  Total horizontal deviation = Horizontal deviation of build + Horizontal deviation of Hold section + Horizontal deviation of drop
- Total horizontal deviation $= r_1 \cdot (1-\cos\Theta) + L_2 \cdot \sin\Theta + r_2 \cdot (1-\cos\Theta_2)$
- Lead angle [17]

  Lead angle is the angle to the left of the target location to compensate for the walking tendency of bit in the right. In the old days, it was normal practice to allow a "lead angle" when kicking off, because roller cone bits used with rotary assemblies tend to "walk to the right". In extreme cases, the lead angles could be as large as $20°$.

Offshore Drilling and Completion

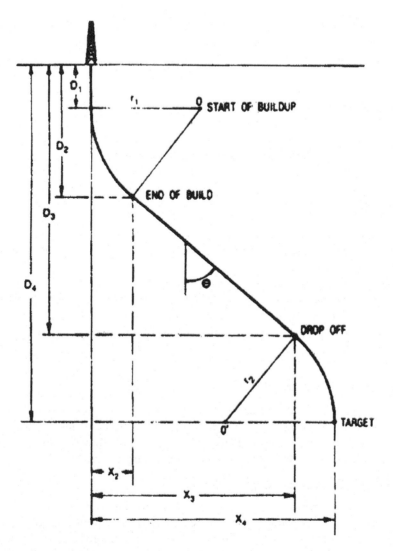

FIGURE 3.13 Well trajectory schematic [16].

Because of greatly increased used of steerable motors and PDC bits for drilling, now there is no need to give lead angle to wells when kicking off.
- **Nudging**

The technique of "nudging" is used on platforms to "spread out" conductors and surface casings, which minimizes the chance of a collision.

Other reasons for "nudging" include:
- To drill from a slot located on the opposite side of the platform from the target when there are other wells in between.
- To keep wells drilled in the same general direction as far apart as possible.

- If the required horizontal displacement of a well is large compared to the total vertical depth, then it is necessary to build angle right below the surface conductor to avoid collision.

  The directions in which the wells are "nudged" should be chosen to achieve maximum separation. Wells may not necessarily be nudged in their target directions.
- **Proximity (anti-collision) analysis**

  To eliminate the risk of collisions directly beneath the platform on multi-well projects (particularly offshore), the proposed well path is compared to existing and other proposed wells. The distances between the other wells and the proposal are calculated at frequent intervals in critical sections, using the EC*TRAK software (BHI) or COMPASS.

Survey uncertainty must also be computed for both the proposed well and the existing wells. The minimum acceptable separation of wells are usually linked to "cone of error" or "ellipse of uncertainty" calculations.

All the above-mentioned well trajectories and their terminologies will guide us through directional drilling operations. Directional drilling and extended reach drilling are common operations in offshore operations due to greater water depths. Thus, directional drilling along with emerging drilling technologies is explained in the next section.

3.1.7 DIRECTIONAL DRILLING

Directional drilling has been described as "the art and science involved in the deflection of a wellbore in a specific direction in order to reach a pre-determined objective below the surface of the Earth". Directional drilling is the process of directing the wellbore along some trajectory to a predetermined target.

3.1.7.1 Evolution of Directional Drilling

The oil and gas industry developed directional drilling in the 1920s, and while the concept has remained the same, the technology has improved greatly over the years. The origins of directional drilling in the oil industry go back to the late 19th century in the United States. Rotary drilling techniques are being introduced, replacing the older cable-tool rigs. The oil and gas industry have developed a number of technologies to improve overall efficiency, such as using advanced drill sensors and global positioning technology to ensure the success of a well. In addition to these technologies, other tools such as whip stocks, bottom-hole assembly configurations, three-dimensional measuring devices, as well as specialized drill bits and motors have enabled a single location to service multiple wells dug at nearly any angle, thus tapping reserves more than a mile deep and miles wide.

3.1.7.2 Types of Directional Wells

1. Slant
2. Build and hole

Offshore Drilling and Completion 65

3. S-curve
4. Extended reach
5. Horizontal

3.1.7.3 Conventional Drilling versus Directional Drilling

Drilling in deviated holes is similar to drilling in vertical holes with allowances made for the tools and hole deviation. Bits generally are similar, sometimes with more side-cutting action during deviation. Drilling is slower because most directional assemblies cannot operate efficiently at the higher bit weights and rotary speeds common to vertical drilling. This reduces the penetration rate correspondingly. There are more non-drilling-type operations, resulting in an increase in total time. These include measurements, orientation, longer circulation periods for hole cleaning, extra trips for various assemblies, and slower trip time due to extra drag and torque. This non-drilling time should be kept to a minimum. There also are more drilling and related problems, especially in holes with higher inclinations and more complex patterns. The net result is that less time is spent drilling; directional drilling is slower; and has more risk of failures. Careful planning is essential. For example, making a correction with a deviation assembly after drilling with rotary assemblies takes one or two days and can be eliminated by predicting bit walk accurately [14].

To understand directional drilling technology it is necessary to understand the different directional drilling tools.

3.1.7.4 Directional Drilling Tools

- Drilling tools
- Surveying/Orientation services
- Steering tools
- Conventional rotary drilling assemblies
- Steerable motors
- Instrumented motors for geosteering applications
- Rotary steerable systems
- At-bit inclination sensor

3.1.7.5 Measurement While Drilling (MWD) Tool [18]

For wireline logging, we had to stop the drilling process, put the drill pipe on slip, break out Kelly, lower the wireline tool, retrieve the tool, read the survey, and plan further action. These increase the non-productive time (NPT), and in offshore, as the day rates are much higher than the onshore rigs, NPT results in huge amount of loss to the company. Therefore, to supplement this issue, MWD and LWD are preferred in offshore directional drilling. The MWD tool transmitted the survey reading to surface through the mud stream in the drill pipe. The drilling process was stopped for few minutes and survey readings were obtained in pump off condition. This saved time to a greater extent compared to wireline logging. The transmission of survey data though mud stream was one of the means. Thus, MWD is considered

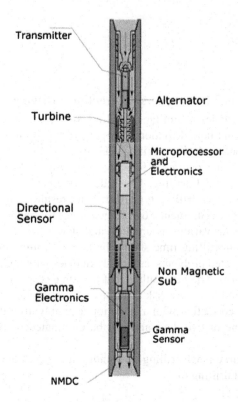

FIGURE 3.14 MWD tool [18].

a better option for survey data transmission compared to wireline procedure. MWD was equipped with gamma ray sensor to detect the natural radioactivity and characterize shale presence, gauge to measure annular pressure which are useful in slim hole to determine ECD, and strain gauge to measure weight on bit (WOB) and torque on bit (Figure 3.14).

MWD tools can also provide information about the conditions at the drill bit, including:

- Rotational speed of the drill string
- Smoothness of that rotation
- Type and severity of any vibration downhole
- Downhole temperature
- Torque and weight on bit, measured near the drill bit
- Mud flow volume

3.1.7.6 Importance and Uses of Directional Drilling [14]
- Directional drilling has now become an essential element in oilfield development, both onshore and offshore.

Offshore Drilling and Completion

- With horizontal drilling, productivity of the same formation is increased up to 20 times than conventional drilling.
- With the addition of high-temperature-resistant drilling motors and modified drill bits, which can drill into nearly any surface, directional drilling is now possible in nearly any geological environment.

3.1.7.7 Application of Directional Drilling

1. Multiple wells from single location:
   - An optimum number of wells can be drilled from a single platform or artificial island. This greatly simplifies gathering systems and production techniques.
2. Inaccessible locations:
   - If a reservoir is located under river beds, mountains, cities, etc., this technique of directional drilling is used.
3. Fault drilling:
   - This eliminates the hazard of drilling a vertical well through steeply inclined fault plane, which could slip and shear the casing.
4. Salt dome drilling:
   - To reach the producing formation, which often lies underneath the overhanging cap of the dome, the well is first drilled at one side of the dome and is then deviated to producing zone to avoid drilling problems such as large washouts, lost circulation, and corrosion.
5. Side tracking and straightening:
   - It is used as remedial operation either to side track obstruction by deviating the wellbore away from obstruction by deviating the wellbore back to vertical by straightening out crooked holes.
6. Relief well drilling:
   - This technique is applied to the drilling of relief wells so that heavy mud may be pumped into the reservoir of the uncontrolled well.
7. Controlling vertical wells:
   Horizontal well:
   - Reduced production in a field may be due to many factors, including gas and water coning or formations, with good but vertical permeability. Engineers can then plan and drill a horizontal drain hole. This is a special type of directional well.
   - Horizontal wells are divided into long, medium, and short-radius designs based on the build-up rates used. Other applications of directional drilling are in developing geothermal fields and in mining.
   - ERD well:
     - Replace subsea wells and tap offshore reservoirs from fewer platforms
     - Develop near shore fields from onshore
     - Reduce environmental
     - Impact by developing fields from pads (Figure 3.15)

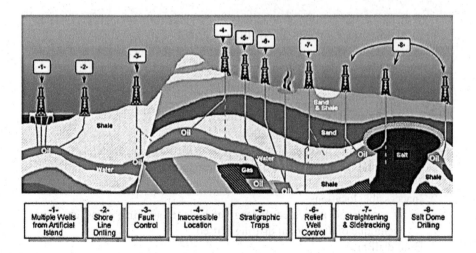

FIGURE 3.15 Directional well drilling application [14].

3.1.8 Dual Gradient Drilling

3.1.8.1 Introduction

Drilling for hydrocarbons offshore, in some instances hundreds of miles away from the nearest landmass, poses a number of challenges over drilling onshore. To meet the world's increasing demand for energy, deepwater exploration and production has become a requisite in the past years. Hence, the need for newer technologies becomes essential for safe and successful operation into deep water. The wellbore pressure (at any depth) should be maintained between the naturally occurring pressure of the formation fluids and the maximum wellbore pressure the formation can maintain without fracture. A variety of new drilling technologies are being developed to address the phenomena of narrow pore-pressure/fracture-gradient windows as pore-pressure/fracture-gradient prospects are drilled in greater water depths. Among them, one of the more proficient drilling types is the dual gradient drilling (DGD). We will discuss in detail the concept of DGD, its significance in deepwater offshore drilling, history of origin, the system description, types of DGD, advantages and limitations with an overall review of DGD process.

3.1.8.2 Dual Gradient Drilling – Overview [19]

In DGD, the pressure profile in the annulus has two distinct pressure gradients. An example would be a heavy mud below the midline and a seawater gradient above the mud line.

3.1.8.3 Single versus Dual Gradient Drilling

See Figures 3.16 and 3.17.

Offshore Drilling and Completion

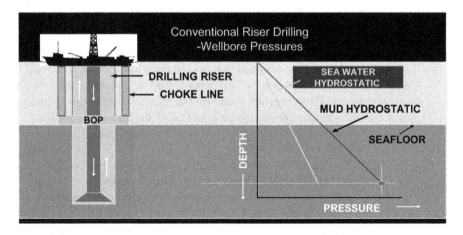

FIGURE 3.16 Conventional riser drilling.

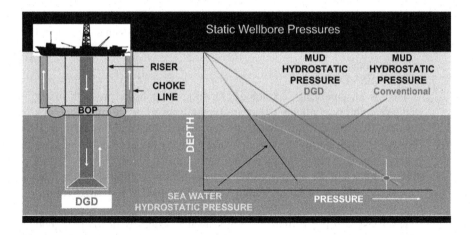

FIGURE 3.17 Dual gradient drilling [19].

3.1.8.4 Dual Gradient System

The subsea equipment of DGD includes:

- Marine riser
- Subsea rotating diverter (SRD)
- Drill string valve (DSV)
- Mud lift pump (MLP)
- Return lines
- Mani folding
- Dual trip tanks

70 Offshore Operations and Engineering

3.1.8.5 Types of Dual Gradient Drilling [19]

1. PUMP-BASED DGD

   Pump-based DGD (PGDB) is where all the systems rely on subsea pumps. The drilling systems were designed to provide a continuous loop circulating system for pumping mud down the drill string and up the annulus to the wellhead, where the wellbore pressure at the seafloor is essentially the hydrostatic pressure of the seawater column above it. In PGDB, the hydrostatic pressure in the wellbore at the wellhead is at or near seawater hydrostatic.

2. DILUTION-BASED DGD

   This employs two fluid gradients in the wellbore. In dilution-based DGD, the "riser" mud weight is determined by the "downhole" mud weight (the density of the mud pumped down the drill string and up the annulus to the injection point, nominally just above the BOP stack), the density of the "dilution" mud (which is pumped down an auxiliary line such as the riser boost line), and the ratio of their respective flow rates (the "dilution ratio"). These different densities and dilution ratios are engineered to meet the particular requirements of the hole section being drilled. A bank of specialty centrifuges is installed to separate the returning "riser mud" back into the heavy "downhole" and light "dilution" mud weight components. Additional control equipment is also required to monitor and control fluid densities and flow rates.

3. RISER-LESS DGD

   This is an emerging technology. Drilling deep holes presents major wellbore stability challenges that are typically mitigated through the circulation of dense drilling mud to prevent hole collapse and remove drill cuttings. This is normally accomplished through the application of riser system. However, riser lengths are presently limited to use in water depths of around 3,100 m. Drilling in extreme water depths requires the use of the "riser-less" drilling technique which is not constrained by the length limitations of a riser system.

   Riser-less drilling primarily uses seawater, rather than drilling mud to manage the borehole because the drilling mud cannot be recirculated. Seawater has lower density than drilling mud, thus, the boreholes are likely to collapse with increasing depth as the pore pressure exceeds the hydrostatic pressure exerted by seawater. For deeper drilling, drilling mud with regulated density must be continuously circulated to keep the borehole from collapsing. The use of alternate techniques, such as riser-less mud recovery (RMR) and DGD, provides an attractive solution to drill deep holes in deepwater settings (Figures 3.18 and 3.19).

RMR provides a DGD setup of the well, while capturing the drilling fluid and returning it to the drillship. The term "dual gradient" implies two hydrostatic gradients: (1) the seawater gradient that begins at the sea surface, and (2) the drilling mud gradient that begins at the seafloor. Conventional drilling has only one pressure gradient for both seawater and mud that originates at the sea surface. Because DGD has much less hydrostatic head associated with the drilling mud in the borehole, drilling fluids

Offshore Drilling and Completion

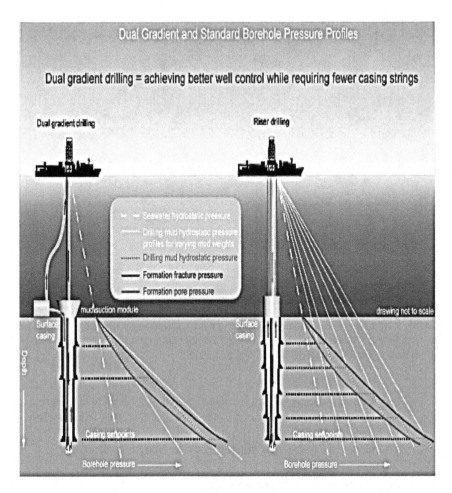

FIGURE 3.18 The dual gradient drilling approach establishes the drilling mud hydrostatic gradient at the seafloor, which allows for the use of heavier drilling muds with an advantageous pressure increase with depth [19].

can be properly weighted, allowing drilling to be more easily contained within the window between the formation pore pressure and the formation fracture pressure, thereby avoiding wellbore instability [17].

3.1.8.6 Limitations of DGD

Problems of DGD faced by operating companies that employed DGD process in the past includes:

- Difficulty in primary subsea cuttings separation that necessities the use electrical submersible pumps (ESP).
    - Major drawbacks in gas dilution DGD includes:
        - Corrosion and inflammability associated with air injection.

FIGURE 3.19 The AGR riserless mud recovery subsea equipment consists of a mud suction module and seafloor mud pumps. The mud is collected in the mud suction module and returned to the drillship via the seafloor mud pumps [19].

- Deeper water depths require huge amount of gas for it to be effective.
- Large gas expansion occurs when the gas traverses from shallow to deep.
- Control of hydrostatic pressure for riser margin is difficult.

3.2 OFFSHORE WELL COMPLETION

3.2.1 Introduction

Once the well has been drilled, cemented, and cased the next decision is about the completion of the well by using equipment which will produce maximum amount of oil and gas at minimum expenditure. Completion of a well is the most important phase of a well's life. The production technique, future workover possibilities, well productivity, downhole problems, etc. depend on how the well was completed. Before a completion operation is designed, various factors need to be taken into account:

- Type and volume of produced fluids
- Target zone depth and surrounding environment
- Presence of multiple zones
- Stimulation potential of the well
- Type of artificial lifts (generally gas lift in case of offshore)
- Frequency of workover operations

Offshore Drilling and Completion 73

After considering all the relevant factors, engineers need to choose from various completion systems available, ranging from simple open hole completions to complex multilateral completions. Optimal completions provide solid flow efficiency, data gathering, and flexibility. Generally, completions designs are planned before drilling but engineers should be prepared for unexpected contingencies which can arise in their plan.

This section discusses various types of completion systems available, equipment utilized, and the safety features specific to offshore operations.

3.2.2 Well Completion Concepts in Offshore

In offshore, completion is referred to in two contexts, a well completion or a subsea completion. Well completions refers to development of a wellbore to obtain controlled recovery of fluids, whereas subsea completion refers to a system of pipes, connections, and valves that reside on the ocean bottom and serve to gather hydrocarbons produced from individually completed wells and direct them to a storage and offloading facility that might be either offshore or onshore.

Other than some specialized and innovative completion solutions, offshore completion and onshore completion are not very different. Most offshore wells are directional and multilateral; completion techniques utilized in these types of requirements are discussed further in the chapter.

Although basic structure of wellheads and trees is also similar to those used on land, these are designed to be compact and lightweight and possess a greater degree of robust remote control so that they can be operated remotely in case of emergencies.

3.2.2.1 Classification of Completions

3.2.2.1.1 Based on Configuration of Casing

- Open hole completions [20]

    In these, the production casing is set on top of or slightly into the pay zone and cemented. The pay zone is left open without cementing. In a typical open hole completion, casing is set prior to drilling into the producing interval. A non-damaging drilling fluid can then be used to drill into the pay section.

This type of completions has several advantages such as:

- Higher production as the flow comes from full wellbore diameter area.
- Savings in terms of no perforations costs.
- Flexibility for future completions and deepening of hole

However, these can only be applied in formations having consolidated formations, and controlling the fluid flow also becomes very difficult. In case of gas and water coning, these completions are not effective in providing control. A simple open hole completion is described in Figure 3.20 (Right).

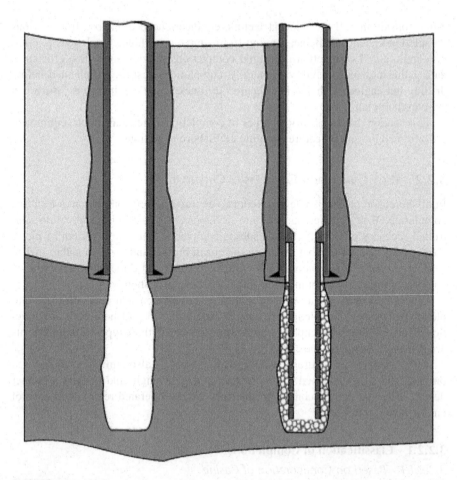

FIGURE 3.20 Left: a simple open hole completion; Right: a perforated liner completion [6].

In open hole completions, production casing is set on top of the pay zone similar to open hole completions, but the production interval is completed by using a liner. Several types of liner completions are commonly employed in well completions, including:

- Slotted liner (Figure 3.20 Right)
- Screen and liner
- Cemented liner (Figure 3.20 Right)

The slotted liner completion is similar to an open hole completion and has all the major advantages and disadvantages discussed for open hole completions. The only difference is that a slotted liner is hung in the open hole interval to minimize sloughing of the formation into the wellbore and increase the level of integrity a completions provides.

Offshore Drilling and Completion

A screen and liner completion is similar to the slotted liner completion in that a screen and liner is set in the open hole section of the wellbore. The difference is that gravel is sometimes placed behind the screen. The advantages and disadvantages are the same as those for open hole completions. The screen and liner completion is used primarily in unconsolidated formations to prevent the movement of formation materials into the wellbore, restricting the flow of reservoir fluids.

The cemented liner completion is used when intermediate casing is set in a well prior to reaching total depth. Many times intermediate casing is used to isolate zones behind pipe, such as low-pressured intervals, that tend to cause lost circulation problems or isolate zones such as shale or salt layers. Intermediate casing is also set in transition zones between normally pressured intervals and geopressured intervals. After the casing is set, the weight or chemistry of the drilling fluid can be changed to continue drilling the well. The cemented liner completion is advantageous because particular intervals behind the liner can be selectively perforated. This selection will allow one to control both the production and injection of fluids in those intervals. The main disadvantage of a cemented liner is the difficulty encountered in obtaining a good primary cement job across the liner.

3.2.2.1.1.1 Perforated Casing Completions In these completions, the producing interval is covered by the production casing which is cemented in place and then perforated for production. These are the most commonly used completion techniques today. Modern perforating charges and techniques are designed to provide a clear perforation tunnel through the damaged zone surrounding the wellbore. This provides access to the undamaged formation, allowing efficient reservoir drainage.

The technique has several advantages:

- The tubing controls the internal corrosion of the casing because produced fluid flows through it and does not contact the casing.
- A good cement job can be achieved and zonal isolation is successful.
- Finally, the perforated casing completion is quite adaptable to multiple completions and alternate completions.

However, the tubing restricts the flow of produced fluid, and the completion is more expensive because of the cost of the packer, tubing, and accessories (Figure 3.21).

3.2.2.1.2 Based on Number of Production Strings or Number of Zones

3.2.2.1.2.1 Single Zone Completion When a single zone has to be completed, a single tubing string is lowered into the well to take production from a single layer, as shown in Figure 3.20.

Economics of production and functional requirement are the driving force for determining the nature of completion. Single zone completion is the easiest completion as one string with a packer or isolation device is used to provide protection to both the casing and liner. A flow control device is also installed.

3.2.2.1.2.2 Multiple Zone Completion In case production is taken from multiple layers through a single string, then the completion string uses packers and

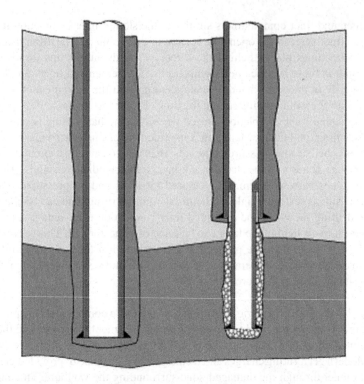

FIGURE 3.21 Left: casing completion; Right: cemented liner completion [6].

sliding sleeves to control flow from individual layers. It is important that the reservoir pressures of all the layers flowing into a single string be similar for this type of completion to be successful.

Simultaneous production from multiple zones verses selective production from selected zones, depending on reservoir requirement along with other functional requirement and economics of production, decides the selection of configuration of multiple zone completion. ***Dual-zone single string completion*** (Figure 3.22)

The flow from two layers is taken through tubing and casing. This type of completion is not used in offshore as production through casing and tubing annulus is considered a safety hazard. Such a configuration does not allow upper formation to be produced through tubing.

3.2.2.1.2.3 Dual-Zone Dual Packer Completion In this type of completion, there is flow in tubing and annulus. However, it allows the upper zone to be flowed through the tubing. Again, for safety of casing, such completions are not used in offshore unless where there is provision to divert the production through annulus to tubing string.

3.2.2.1.2.4 Dual-Zone Parallel String Completion (Figure 3.23) In this type of completion, two layers are completed and production from each layer is taken through different production strings. Such a type of completion does not require the

Offshore Drilling and Completion

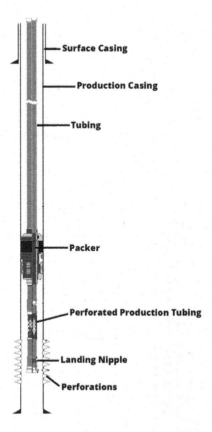

FIGURE 3.22 Dual-zone single string completion. (Courtesy: Baker Hughes.)

pressures of individual layers to be similar as production from each layer is independent of one another. The layers are isolated through the use of packers.

3.2.2.1.3 Advantages
- It is possible to produce from/inject into more than one production/injection zone through a single well, thereby reducing the overall development costs.
- Selective treatment of individual zone is possible.
- Use of natural energy from one zone can be used to artificially produce another zone.

3.2.2.1.4 Disadvantages
- Large number of equipment downhole used can create problems.
- Expensive and more complicated completion and workover technique.

There is a possibility of loss of production in zone due to mechanical problems and formation damage during workover.

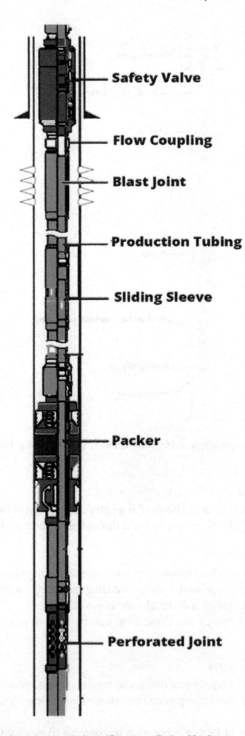

FIGURE 3.23 Single zone completion. (Courtesy: Baker Hughes.)

Offshore Drilling and Completion 79

3.2.2.1.4.1 Multiple Zones Single String Completion (Figure 3.24) In such completions, the producing zones are opened or closed individually through the use of wire line or hydraulic pressures.

3.2.2.1.4.2 Naturally Flowing Completions Natural flow of well is possible when reservoir pressure is higher than the summation of hydrostatic pressure exerted on reservoir due to depth and friction offered in flow path. Well completion for a

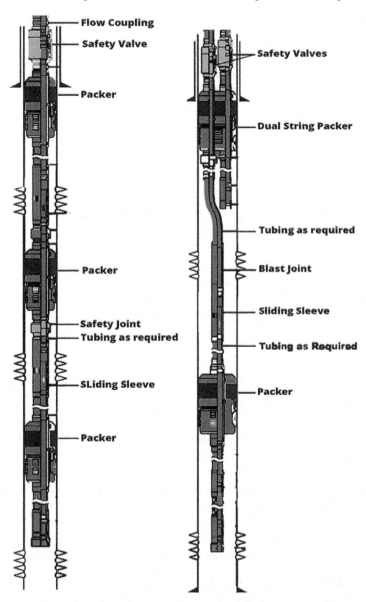

FIGURE 3.24 Dual-zone parallel string completion. (Courtesy: Baker Hughes.)

80 Offshore Operations and Engineering

naturally flowing well is simple and economical. Simpler downhole components and equipment are required. However, completion of high-temperature and high-pressure wells are complicated and costly due to the specific requirements of high-temperature and high-pressure reservoir and its safety requirement.

During well completion design, requirement of other production methods, like artificial lifts, should be considered to avoid design complication at a later date.

3.2.2.1.4.3 Artificial Lift Completion   When production from naturally flowing well starts depleting or reservoir pressure of a new well is not sufficient to support natural flow, artifical lift systems are deployed. Specialized downhole components/equipment consisting of electrically or mechanically operated pumps/valves and/or other precision engineered devices are placed to increase or sustain the production level. These systems have some component/equipement placed on the surface. Periodical or breakdown maintenance is a requirement of such systems involving workover operation.

Artificial lift system:

- Gas lift
- Electric submersible pump
- Plunger lift
- Hydraulic or jet pump
- Variable cavity pump (VCP)
- Progressive cavity pump (PCP)

Of these, gas lift is very commonly used for offshore applications. When using gas lift to enhance production, a packer is utilized to separate the produced fluid pathway from the injected gas pathway down the annulus. Packers are often used with ESPs to facilitate control of well zones. Tubing anchors are commonly used to increase the efficiency of rod pumps. Anti-rotational anchors are commonly used with PCPs. Gas lift and ESP completions are shown in Figures 3.25 and 3.26.

3.2.3 Horizontal Well Completions

3.2.3.1 Open Hole Completion

In open hole completion, the horizontal segment of the wellbore is kept open without any casing or liner. This type of completion is very inexpensive and is well suited for a stable formation that remains stable throughout the life of the well. It completion ideally suits the formations having very little zonal isolation requirement for water and gas breakthrough. The most common application for open hole completion is in fractured limestone or chalk formations. Another advantage is that the formation does not get damages due to completion, for example, cementing, etc. Absence of a casing in open hole, however, severely limits stimulation and remedial options, including control either in injection or production throughout the life of the well.

3.2.3.2 Slotted Liner Completion

The only difference between open hole and a slotted liner completion is the addition of a slotted liner having narrow longitudinal slots/holes in the open hole

Offshore Drilling and Completion

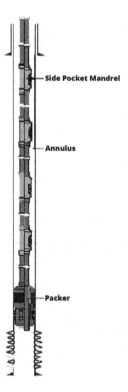

FIGURE 3.25 Multiple-zone single string completion.

section of the well. Convenient path for entry of downhole tools during workover operations is an added benefit of slotted liner apart from additional stability to the open hole section without adding substantial cost or complexity over the open hole completion.

Sand control in unconsolidated sandstone reservoirs is achieved to considerable degree by pre-pack/gravel pack liners.

The conventional slotted liner completion fails to provide any zonal isolation, which makes selective layer simulation nearly impossible. This is the major disadvantage of conventional slotted liner completion.

3.2.3.3 Slotted Liner Completion for Zonal Isolation

Selective zonal isolation and stimulation to a great extent can be achieved by deploying ECPs during completion. This is an improvement over conventional slotted liner completion.

A long horizontal section is divided into several producing section by installing inflatable ECPs outside the slotted liner in open hole. ECPs are used to isolate unwanted horizontal section from producing zone. Ported subs or sliding sleeves, bridge plugs, packers, or straddles can be used for achieving limited zonal isolation.

Production control along the wellbore with limited zone isolation can be achieved by this type of completion at lesser cost and risk than cemented and perforated

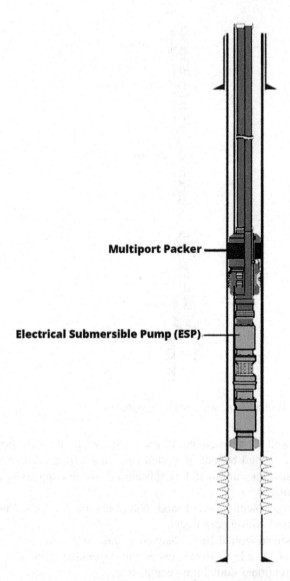

FIGURE 3.26 Left: A gas lift completion; Right: ESP completion. (Courtesy: Baker Hughes.)

completion. This type of completion is not favored for reservoirs with corrosive and scaling environment due to risk of metal corrosion and scale deposition.

3.2.3.4 Cemented and Perforated Completion

After lowering of casing or liner in wellbore section, cement is pumped at adequate pressure to form a bond between casing/liner and wellbore. Perforation is done after completion of cementation in hydrocarbon-bearing section. This completion is known as cemented and perforated completion. Retrievable or permanent plugs or

Offshore Drilling and Completion

packers are used for achieving effective zonal isolation. This type of completion is normally adopted in following conditions:

- Low permeability formation requiring either transverse or longitudinal hydraulic fracture to ensure economical drainage for production.
- Reservoir with top gas or lower water layers having thin oil column necessitating frequent well operation for isolation/shutting-off unwanted gas and water production zone.

3.2.4 INTELLIGENT WELL SYSTEMS

Wells are completed with component and facilities in place to collect flow, pressure, and temperature data at different levels of interest in the well to facilitate online monitoring and analyze well behavior and production. Wells completed in line with the above-mentioned requirements are called intelligent or smart wells. Some of the significant capabilities available as a result of such completion include:

1. Collection of real-time data of reservoir, as well as wellbore and wellhead parameters
2. Real-time monitoring of production and well
3. Capability of remotely controlling the well

Intelligent well system is the foundation stone for creating an intelligent producing field. These systems have the following key benefits.

- Best possible optimization of hydrocarbon recovery with reduction in water cut.
- Real-time data monitoring and analysis to achieve optimized infill development targets.
- Better understanding of reservoir for oil and gas production
- Optimized utilization of artificial lift/secondary or other recovery systems.
- Eliminating unwanted well intervention, water handling, etc., thereby reducing operating costs.
- Controlled co-mingling of zones, as well as monitoring and control of multiple laterals.
- Multiple simultaneous functions from the same well such as utilizing the same well for injection well, observation well, and production well.
- Monitor and manage well integrity.
- Monitor environmental conditions.

All the above help reduce the capital cost for the complete project. A simple schematic is shown in Figure 3.27.

3.2.5 MULTILATERAL COMPLETIONS

In view of the rising cost of hydrocarbon exploitation, multilateral well completion systems have become a convenient tool to reduce cost by utilizing single slot or

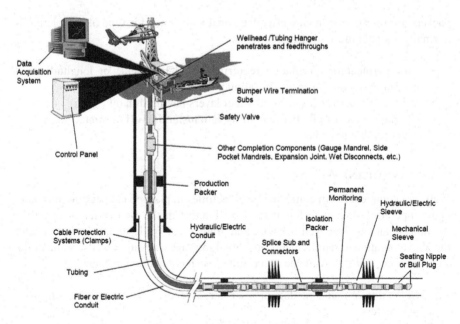

FIGURE 3.27 A schematic of an intelligent well system with various components. (Courtesy: Baker Hughes).

mother bore for producing from multiple layers. Type of multilateral completion depends on reservoir characteristics, production profile, lifting mechanism, sand control/debris control, water handling, etc.

3.2.5.1 Multilateral Well Classification

Type of completion for multilateral well defines various multilateral well types. Drilling operations do not vary much for different types of multilateral wells. However, the completion hardware used in the two systems vary widely, and the risk involved is also different.

An oil industry forum on the Technical Advancement of Multilaterals (TAML) has been created to develop multilateral classification matrix and foster a better understanding of multilateral well application, capabilities, and equipment. Major global oil and service companies are the members of this forum.

To properly categorize various multilateral systems, TAML group has classified the systems into levels as a function of increasing risk and complexity. In this, a well's classification level corresponds to the highest junction level in the well (Figure 3.28).

3.2.5.2 Level 1 Multilateral Well

This type of multilateral well is drilled from an open hole main bore (parent bore) without any mechanical support or hydraulic isolation at the junction. These are the simplest multilateral systems and utilize multiple drainage legs from a parent open hole bore. The advantages of this system are its low cost and simplicity. There is no milling required, no whipstocks to retrieve, no liners to cement, and no production

Offshore Drilling and Completion

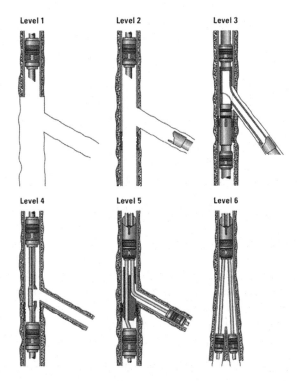

FIGURE 3.28 TAML multilateral well classification. (Multilateral wells, Oilfield Review 2000, Schlumberger.)

control equipment to install. Lack of casing support limits its use to stable formation only at the cost of no production control of individual layers. The production must be commingled and cannot be selectively shut-off.

3.2.5.3 Level 2 Multilateral Well

Level 2 multilateral system has a cased parent bore and open or simple (slotted liners, pre-packed screens) lateral bores. The parent bore is cased and cemented, and an orientation packer is set below the proposed junction kick-off point. A whipstock is then landed into the packer and oriented in the desired kick-off direction. Subsequently, the casing is milled away and a lateral is drilled to the target length. After drilling the lateral section, the whipstock is retrieved and the hole is completed in various ways with different completion hardware depending on well requirements.

The requirement of casing exit and whipstock retrieval in Level 2 poses a significantly greater risk than Level 1, but minimizes the chances of borehole collapse and provides hydraulic isolation between the lateral zones.

A low cost option of multilateral completion system that provides both parent bore support and production control, not found in Level 1, is to incorporate a sliding sleeve between the orientation packer and a second packer placed above the casing exit so that production from both the laterals can be commingled. Production isolation of either lateral is possible in this arrangement. In case bottom zone begins to deplete or produce water, a plug can be placed in the lower packer to shut off production from the lower bore. In case the upper lateral production needs to be shut off, then the same can be done by closing the sliding sleeves.

However, the main disadvantages of this type of completion system are continued reliance on formation for junction stability at exit point, no provision to produce separately from individual layers, and limitation of re-entry in lateral bore due to sliding sleeve. Re-entry into lateral bore is possible in case a lateral nipple is used instead of a sliding sleeve.

3.2.5.4 Level 3 Well Completions

The next level of completion in terms of risk and complexity is Level 3 that has its main bore cased and cemented and laterals cased but not cemented. Lateral liner is anchored with main bore but not cemented. The well can be completed with a lateral re-entry system or a sliding sleeve in the same manner as in Level 2 completion. Quality of mechanical support at the junction is achieved by anchoring the lateral liner, however, still hydraulic support is not available at the junction. Cement sheath in main bore at the junction is not capable of withstanding even a small pressure differential and can fail over a period of time if draw down becomes substantial. Therefore, this system is not useful for wells with unconsolidated formations expected with higher drawdown during production.

3.2.5.5 Level 4 Well Completions

Level 4 junction offers both a cased and a cemented main bore and lateral. This gives the lateral mechanical support, however, the cement itself does not offer pressure integrity at the junction. There is a potential for failure if the junction is subjected to a pressure drawdown. Zonal isolation and selectivity is possible by installing packers above and below the junction in the main bore [13].

3.2.5.6 Level 5 Well Completions

The Level 5 multilateral junction offers cased main bore and lateral, and provides increased mechanical integrity. Pressure integrity is achieved by using completion to isolate the junction. Both the main bore and the lateral can be accessed independently, and the zones can be produced independently or commingled.

3.2.5.7 Level 6 Well Completions

In Level 6 multilateral system, both mechanical and pressure integrity are achieved using the casing to seal the junction. Cementing the junction, as was done in the Level 4 system, is not acceptable. The Level 6 system uses a premanufactured junction. In one type of system, the junction is reformed downhole. TAML6 junction as of today are not considered to be viable because of significant loss of ID and complex installation. Contemporary TAML5 systems offer higher pressure ratings and IDs, which almost made TAML6 junctions obsolete [11].

3.2.6 Subsea Completion

A subsea completion refers to a system of pipes, connections, and valves that reside on the ocean bottom and serve to gather hydrocarbons produced from individually

Offshore Drilling and Completion

completed wells and direct them to a storage and offloading facility that might be either offshore or onshore. It is different from a platform-based well where tubing carrying hydrocarbon with full pressure continues up to the platform deck. The emergence of subsea technology has revolutionized the industry's offshore activities, and has developed at a remarkable pace in recent years In general, surface completion systems are cheaper to manufacture, easier to install, and far less troublesome to maintain than subsea systems; hence, the decision to opt for subsea development is generally taken when other context-specific criteria make it demonstrably superior in terms of overall cost-effectiveness. Such criteria include water depth, prevailing climate and environmental conditions, well numbers, reservoir size and reserve distribution, well maintenance requirements, etc. [21]. Conditions that particularly favor the adoption of subsea completion technology include:

- Deepwater fields
- Small/Marginal field or fields with scattered reserve distribution
- Harsh environmental conditions
- Fast track development projects
- Phased development – designed to achieve early production that can then be augmented by later stages of development

In some situations, subsea developments linked to floating production system (FPS) or tied to the existing platforms provide the most cost-effective development options. One of the fundamental aspects of field development, including deepwater development, is whether to have an above-surface wellhead or subsea. In deep waters, the choice is limited due to water depths, availability of appropriate technology, and extreme cost limitations in for opting a surface tree. Subsea well completion has proved to be reliable in service and cost, and hence, the numbers of subsea completions have increased over the years, especially for the development of deepwater fields where subsea completion presents a low cost-development solution in deep water. There are, however, several key elements that need to be in place to find an optimal subsea solution. The subsea wellhead may also be designed to accommodate a tieback to a surface facility on a TLP, SPAR, or a fixed platform. This section will provide a brief description of subsea completion, while all the other components and equipment involved in subsea completion will be discussed in detail in Chapter 6.

3.2.6.1 Types of Subsea Completions

a. Individual Satellite Well

Satellite wells consist of individual wells with separate control umbilical and flow line for each well linking production back to the floater or platform. In the past, subsea connection and maintenance work was carried out by divers, and therefore, these systems were depth-limited. Moreover, the cost of individual flow lines and umbilical was prohibitively expensive in case of long distances. However, with technological advances in production hardware and development of remotely operated vehicles (ROV), the satellite concept has become a practical and effective development strategy for deepwater prospects and has been adopted in many current projects.

b. Cluster Development

The cluster concept permits drilling and completion independent of the manifold. This allows for a smaller manifold structure, which is cost-effective and can be made retrievable through a drilling rig moon pool. This results in reduced installation cost and provides more installation options. Piling and leveling of the manifold are less stringent due to the size and overall loading, although soil parameters can be the governing factor. In case deepwater production wells are scattered over a wide area and well maintenance is relatively low, then subsea systems involving seabed completion are selected.

Clustered wells provide inherent flexibility. The clustered approach allows:
- Location of the wells to best develop the reservoir
- Scheduling well drilling and other construction activities to accommodate equipment and installation vessel availability, as well as to meet production needs and investment plans
- Future expandability with minimal up-front investment.

Some of the disadvantages of cluster developments are the costs associated with flow line and umbilical jumpers and the need for a centralized location, which, in turn, necessitates extended-reach drilling requirements (Figure 3.29).

c. Integrated Manifold

Integrated template/Manifolds are used to combine flows from a number of wells. A template can provide the base for multiple wellheads, manifolds, and protective structures. In the early 1970s, designs emerged for integrated

FIGURE 3.29 Cluster development concept. (Offshore Energy Today, May 2012.)

Offshore Drilling and Completion 89

subsea template/manifold through which a dozen or more wells could be drilled and completed allowing the commingling of their production. Typically, in the case of a 12-well development, instead of using individual flow lines and central lines for each well, only five lines will be required in the case of integrated template layout – two production flow lines, one injection flow line (if applicable), and perhaps a test flow line and a single electrical/hydraulic control umbilical.

However, in addition to increased complexity and associated costs, the integrated template/manifold has several significant disadvantages such as its heavy weight, which requires a dedicated crane barge for its installation, increased drilling costs in terms of extended reach drilling from a central location, need for precise levelling at the sea bed, etc. These factors can offset the savings made in terms of flow lines and umbilicals.

3.2.7 COMPLETION EQUIPMENT

3.2.7.1 Christmas Tree (Xmas Tree)

A Christmas tree is the most important well control equipment used in well completions. It consists of an arrangement of valves, spools, flanges, and connections to control the flow of fluids from the well. Based on the application and environment of use, several types and configurations of Christmas trees are available. Primarily Christmas tree used offshore is of two types: (1) Platform completed (Dry tree), and (2) Subsea completion (Wet or dry tree). Platform completed wells (Dry tree) are similar to those being used onshore. However, subsea trees are entirely different and will be discussed separately in Chapter 6.

3.2.7.2 Production Tubing

Production tubing forms the conduit for reservoir fluids to flow from wellbore to surface. In addition, it facilitates wellbore service operations such as wire line, stimulation, and circulation. Typically, tubing is run inside a casing or liner but can also be cemented in slim hole wells as the casing. Depending on the type of completion, one or two tubing strings may be used in the well.

Major considerations in the selection of tubing for a particular well include:

Tubing size is determined on the basis of inflow performance of the reservoir and tubing performance to ensure optimum production rates over the life of the field. Tubing sizes from $2\frac{7}{8}''$ to $5\frac{1}{2}''$ are in use.

Tubing grade determines the chemical composition as well as the physical and mechanical properties of tubing. The tubing grade selected for a particular completion must satisfy the minimum performance requirements for that application. Tubing of a sufficient yield strength to withstand the various forces caused by changes in pressures and temperatures must be used in the well. The tubing must also be resistant to formation fluids containing corrosive components, such as H_2S, CO_2, chlorides, and water. Normally, L-80 tubing grade is used which provides resistance to sulfide stress cracking.

90 Offshore Operations and Engineering

Tubing weight determines the burst and collapse ratings of a tubing and is normally expressed in pounds per foot (ppf) and is a function of wall thickness.

Tubing connections are primarily either API connections or premium connections. The commonly used API connection is EUE that provides reliable service in a majority of wells. Premium tubing connections are used in corrosive environments, high-pressure wells, and in wells with bends and doglegs.

3.2.7.3 Packers

A packer provides a means of sealing the tubing string from the casing or liner, thereby preventing fluid movement between them. This protects the casing from undue stress in the form of pressure differentials, as well as protects the casing against corrosion and erosion from the produced fluids. Because casing used in a well is a permanent component of the completion system, repair/replacement of casing is very complicated and expensive. The packer along with tubing string is easier to remove and replace. Packers are also used for zone separation as in the case of multiple zone completions.

All packers mostly consist of:

i. **Flow mandrel** to provide the flow conduit for production.
ii. **Resilient elements** to ensure the tubing to annulus pressure seals.
iii. **Cone or wedges** to assist in positioning of the slips.
iv. **Slips** to grip the casing wall and prevent the packer from moving up and down.
v. **Hold down buttons** to prevent packer from unseating.

The packer design also provides for a spacer tube that has holes to remove trapped air and bypass ports to circulate out debris settled on packer and pressure equalization across the packer elements.

The criteria for packer selection must consider:

- Selection/completion strategy
- Rig capacity for fishing/milling
- Fishing requirements
- Well fluid characteristics, H_2S, CO_2
- Bottom hole pressure and temperature

Packers can be single, dual, or triple bore, and are mainly classified as:

3.2.7.3.1 Retrievable Packer

A retrievable packer is run as an integral part of the tubing string, and is set either mechanically or hydraulically and can be released by pulling or rotating the string.

It provides advantages of reuse and saves rig time as milling operation is not required for removing the packer. However, these have low differential pressure rating, and pressure distribution is also not uniform. As mechanically set packers are temperature-sensitive, care should be taken while using cold fluids during simulation.

The mechanically set retrievable packers are set by applying sufficient right hand rotations to the string and released by straight pull. Compression-set mechanical packers have dies downwards. Tension-set mechanical packers are used when packer is to be set at shallow depths where required compression cannot be given. Packer configuration in tension-set packer is opposite to that of compression set, that is, slips at top and rubber elements at bottom. Prior to running in the packer, it must be ensured that the rubber elements, spacer rings, dies, slips, and Teflon ring at the top sub are in good condition (Figures 3.30 and 3.31).

The hydraulic-set retrievable packers are set by applying a pressure of 1,400–2,000 psi inside the string, thus requiring pump-out plug (POP) to apply pressure. The POP rating depends on well pressure. The ratchet mechanism in the packer stores the setting pressure and enables the elements to remain in inflated condition. The packer is released through a release ring/screws that shear at an overpull of 20,000 to 30,000 lbs above the pullout weight (Figure 3.32).

3.2.7.3.2 Permanent Packer

Normally, once set, the permanent packer, is regarded as part of the casing and can only be removed destructively by milling. The completion string can be engaged for providing the flow conduit or removed from the packer for well killing. Permanent packers can be set mechanically, hydraulically, or electrically through wire line.

These packer types are recommended for when long-term completion, high-pressure differential, maximum dependability, and large packer bore are required (Figure 3.33).

FIGURE 3.30 Tension-type retrievable packer. (Completions Primer 2000, Schlumberger.)

FIGURE 3.31 Compression-type retrievable packer. (Completions Primer 2000, Schlumberger.)

FIGURE 3.32 Hydraulic-type retrievable packer. (Completions Primer 2000, Schlumberger.)

FIGURE 3.33 Permanent packer. (Completions Primer 2000, Schlumberger.)

FIGURE 3.34 An illustration of an inflatable packer. (Courtesy: Drillingcontractor.org.)

3.2.7.3.3 Inflatable Packer

Inflatable packers are run through the tubing string, either on wire line or coiled tubing, and inflated to the required size. The pressure rating of such packers is less. These packers are used in straddle completions and for open hole testing (Figure 3.34).

3.2.7.4 Blast Joint

Reservoir fluids entering the wellbore through perforations may display a jetting behavior that can erode the tubing string at the point of fluid entry, and ultimately may cause tubing failure. The blast joints are pipe joints of 20–30 ft length with a wall thickness greater than the tubing and are run in the string to be opposite of the perforations. The blast joints delay the erosional failure at the point of entry of fluids into wellbore.

3.2.7.5 Flow Coupling

A flow coupling is a short piece of pipe that has a wall thickness greater than the tubing string. It is used to delay the failure caused by erosion wherever turbulent flow is anticipated such as around a landing nipple or subsurface safety valve. The flow couplings are available in 3–10 ft length, and length of flow coupling for a particular application depends on how quickly turbulent flow is expected to dissipate, as well as the abrasive nature of the fluid.

3.2.7.6 Seating Nipple

Seating nipples are located at various depths in the tubing string. The seating nipples enable various wire line intervention jobs for flow control. Some of these jobs include shutting the well for testing the tubing string, circulation, pressure equalization,

Offshore Drilling and Completion

operation of subsurface safety valve when hydraulic control is lost, and installation of downhole chokes.

3.2.7.7 Landing Nipple

A landing nipple is a short section of thick-walled tubing machined internally to provide a locking profile and at least one packing bore. It provides a profile at a specific point in the completion string to locate, lock and seal subsurface flow controls either through wire line or pump-down method.

3.2.7.8 Expansion Joint

Expansion joints are used to compensate tubing movement caused by temperature and/or pressure changes during treatment or production. These are available in various stroke lengths.

3.2.7.9 Safety Joints

Safety joints are used between the packers in multiple completions and in selective completion using hydraulic single string packers.

The shear pin in safety joint enables stuck tubing to be sheared off. However, because it introduces a weak joint, its use should be restricted, wherever possible.

3.2.7.10 Safety Valves

3.2.7.10.1 Surface Safety Valves (SSVs)

SSV is a hydraulically actuated fail-safe gate valve installed in a Christmas tree for producing or testing oil and gas wells with high flow rates, high pressure, or the presence of H_2S. The SSV is used to quickly shutdown the well upstream in the event of overpressure, failure, leak in downstream equipment, or any other well emergency requiring an immediate shutdown. SSVs are known as Hi Lo valve. Figure 3.35 explains the positions of different valves and their link to the main control room.

3.2.7.10.2 Subsurface Safety Valves (SSSVs)

SSSVs are devices installed in upper wellbore to provide emergency closure of production conduits. Two types of SSSVs are available: (1) surface controlled and (2) subsurface controlled. In each case, they are designed fail-safe so that wellbore can be isolated in an event of system failure (Figure 3.36).

3.2.7.10.2.1 Subsurface Control Safety Valves (SSCSVs) Subsurface control is executed by a number of devices such as safety valves, bottom hole chokes and regulators, and injection safety valves. These are installed below the surface, and

FIGURE 3.35 Blast joint (Completions Primer 2000, Schlumberger.)

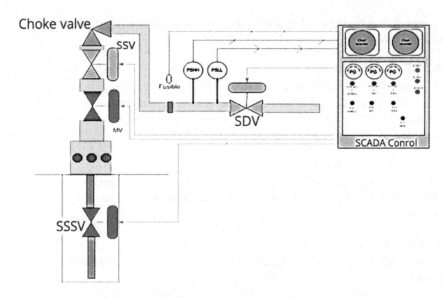

FIGURE 3.36 A schematic of safety valves and control [22].

sometimes are referred to as storm chokes. Based on operating/activation mechanism, these valves can be classified as below.

Differential pressure or velocity type where the valve is spring operated and kept normally open. The differential pressure design valve shuts in the well automatically when there is abnormal production rate; this is caused by rupture in surface equipment resulting in higher pressure differential than the spring setting of the valve.

Pressure-activated type consists of a dome-pressure-operated valve which is normally closed until acted upon by a pressure greater than the preset dome pressure. The pressure-actuated values are precharged with a set dome pressure and held open by well pressure. When the flowing pressure of the well drops, the valve closes to shut in the well. When the tubing pressure is equalized with dome pressure, it will open the valve automatically (Figure 3.37).

3.2.7.10.2.2 Surface Controlled Subsurface Safety Valves (SCSSSVs) The use of SCSSV is mandatory for offshore producing wells and offers a mechanism for remotely operated subsurface well control. It consists of a flapper-type valve located in the string at a depth of 150 m from the wellhead. It is controlled from the surface by hydraulic pressure application through external ¼″ stainless steel control line. The valve has a fail-safe close and is held open against spring pressure by maintaining hydraulic pressure. Loss of hydraulic pressure causes the valve to close and shut in the well. Surface control units that supply the hydraulic pressure also monitor any abnormal increase or decrease in flow-line pressure. The valve operation is independent of tubing pressure and well fluid surges. Both wire line retrievable and tubing retrievable designs of SCSSV are available.

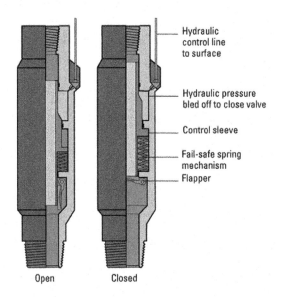

FIGURE 3.37 Subsurface safety valve. (Oilfield glossary.)

3.2.7.10.3 Surface Safety Valves (SSVs)

SSVs are pneumatically controlled and installed in the flow line after flow valve. These are linked to remote operation system either through SCADA system or other communication systems to enable remote operation from the control room located at the process platform or the main control room on land. SSV is remotely operated by an ESD, which can be triggered automatically by high or low-pressure pilot actuators.

3.2.7.11 Circulating Valves

Tubing to annulus communication is required to circulate fluids in a well, treat a well with chemicals, inject fluids from the annulus into the tubing string, or produce a zone that is isolated between two packers. Such tubing to annulus access is provided in the completion string through various circulating devices.

3.2.7.11.1 Sliding Sleeve

Sliding sleeves are the principle circulating devices that provide the ability to circulate a well and selectively produce multiple reservoirs.

A sliding sleeve is a cylindrical device with an inner sleeve and outer body bored to provide matching openings. The inner sleeve is moved using a wire line shifting tool. When the sleeve is moved and matched with openings in the outer body, it creates a circulation path between tubing and annulus. Some of the typical applications for which the sliding sleeves are used are for displacing fluid, selective testing, treating or production in multiple completion, killing by circulation, pressure equalizing, etc. (Figure 3.38).

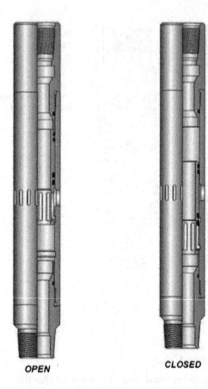

FIGURE 3.38 Sliding sleeve (Open and closed). (Completions Primer 2000, Schlumberger.)

FIGURE 3.39 Side pocket mandrel. (Completions Primer 2000, Schlumberger.)

3.2.7.11.2 Side Pocket Mandrel

Side pocket mandrel has a polished receptacle/pocket on one side that can accommodate downhole tools lowered by wire line. Side pocket mandrels are placed in the tubing string at a location where ever required (Figure 3.39).

REFERENCES

1. *"Basic Drilling Technology," Petroskills.* Available at www.petroskills.com/course/basic-drilling-technology-bdt.
2. *drilling-flash-cards.* Available at https://quizlet.com/26578790/drilling-flash-cards/
3. *"Well design and engineering," Petroskills.* Available at www.petroskills.com/course/well-design-and-engineering-wde
4. *0ldfinal-report.* Available at www.slideshare.net/SureshSanapathi/0ldfinal-report
5. *Drillingfluid.* Available at https://drillingfluid.org/drilling-fluids-handbook/1.html
6. H. Rabia, *Well Engineering & Construction*, Entrac Consulting, London (2002).

Offshore Drilling and Completion

7. M.B. Oyeneyin, *Total Sand Management Solution for Guaranteed Flow Assurance in Subsea Development* (2014).
8. P.P. Benham and F.V. Warnock, *Mechanics of Solids and Structures*, Pitman Publication (1973).
9. *Predicting wellbore stabilty Petrowiki, Jan 2016*. Available at http://petrowiki.org/Predicting_wellbore_stability
10. R.F. Mitchell, *Petroleum Engineering Handbook, Volume II: Drilling Engineering,* (2007) Society of Petroleum Engineers, ISBN: 978-155563-332-5, Editor-in-Chief: Larry W. Lake.
11. R.R.E. Fjær, R.M. Holt, P. Horsrud, and A.M. Raaen, *Petroleum Related Rock Mechanics,* 2nd Edition, Elsevier Science, Burlington (2008).
12. *wellbore instability*. Available at http://petrowiki.org/File:Devol2_1102final_Page_061_Image_0001.png
13. C. Hogg, Comparison of multilateral completion scenarios and their application, in Offshore Europe (1997).
14. *Directional, Horizontal, and Multilateral Drilling - DHD*. Available at www.petroskills.com/course/directional-horizontal-and-multilateral-drilling-dhd
15. *Introduction-to-casing*. Available at www.drillingcourse.com/2015/12/introduction-to-casing.html
16. A.T. Bourgoyne Jr., K.K. Millheim, M.E. Chenevert, and F.S. Young Jr., *Applied Drilling Engineering (SPE Textbook Series 2)*, Society of Petroleum Engineers (1986).
17. *"Offshore Drilling Operations"*. Available at www.petroskills.com/course/offshore-drilling-operations-odo
18. *Measurement While Drilling MWD*.
19. *Dual-gradient system evaluation highlights key high-risk issues*. Available at www.drillingcontractor.org/dual-gradient-system-evaluation-highlights-key-high-risk-issues-17111.
20. *Types of Completions," AAPG Wiki, January 2016*. Available at http://wiki.aapg.org/Types_of_completions
21. B.J. Blythe, L.P. Brzuzy, C.E. Campbell, and J.E. Cooper, *Subsea Drilling, Well Operations and Completions* (2011), pp. 1–45.
22. Annual Report, *Oil and Gas Production Safety System Events* (2017).

4 Offshore Oil and Gas Production and Transportation

Offshore oil and gas production and transportation are crucial and challenging due to limited available space, infrastructure, and sometimes extreme weather conditions. Thus, a detailed explanation is provided in this chapter for the readers to have a thorough understanding of these challenging operations.

4.1 OFFSHORE PRODUCTION OPERATION

As discussed in Chapter 2, a chain of platforms are located offshore to accommodate facilities such as unmanned well (only wells-W) Platform to facilitate production from wells, production testing, and data transmission with the help of remote terminal units (RTU) and telemetry. Process platform to receive produced fluid from Unmanned Well platforms and form other platforms along with well data for further processing and transporting of partially stabilized fluid through pipelines to onshore, where process platform plays the key role. Till pipeline comes, stablised crude can be transported to onshore terminal through tanker. A typical flow diagram depicting the various elements of a production system located on small-to-medium and large platforms is shown in Figure 4.1.

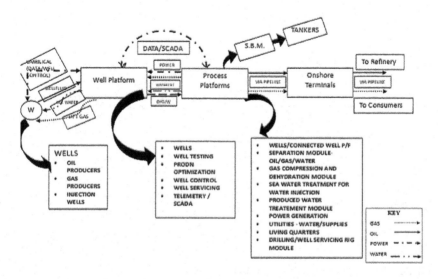

FIGURE 4.1 Overview of offshore production operation.

Process platform is the nerve center of all controls; however, monitoring and emergency control can be also exercised from land-based control through the SCADA system. With advancement in technologies (smart and intelligent well concept), the traditional well platform, with common testing facilities, RTU, and telemetry for data transmission, is slowly becoming obsolete. The major elements of offshore production systems are explained in the below section.

4.1.1 Major Elements of Offshore Production System

4.1.1.1 Wells (Subsea/Platform Wells)

Wells are the first element of the production system, and can be either platform completed (dry wells) or subsea completed (wet wells).

4.1.1.2 Platform Wells/Dry Trees

A dry tree has its wellbore extended vertically up all the way to the host platform. The tree with its various valves and chokes is placed on the platform. The wellbore can be directly accessed from the top for interventions. Platform completed wells are all dry, easy to maintain, and workover in such wells provides comfortable operation.

4.1.1.3 Subsea Wells/Wet Trees

A subsea well is one in which the wellhead – Christmas tree and production – control equipment is located on the seabed. Subsea oil production systems can consist of either single satellite well flowing to a fixed platform, FPSO, or an onshore installation through a flowline, or multiple wells on a template/clustered around a manifold flowing to a fixed or floating facility, or directly to an onshore installation through a pipeline (Figure 4.2).

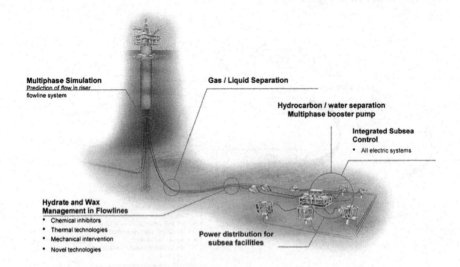

FIGURE 4.2 A subsea field development [1].

Offshore Production and Transportation

Any reservoir, or part thereof, can be developed by connecting subsea drilled wells from one or multiple locations to its dedicated subsea production system. Deployment of subsea production system becomes essential in the case of deepwater conditions or ultradeep water conditions due to technical or economical unfeasibility/limitations of traditional offshore surface facilities, such as on a steel-piled jacket on account of increased water depth.

4.1.1.4 Offshore Pipelines

Pipeline either in offshore or onshore is the basic facility for transporting any liquid from one place to another safely and economically with minimum adverse ecological impact. Different materials starting from ferrous to HDPE can be used for laying of pipeline, depending on transported liquid and its temperature and pressure. Pipeline laid in offshore is known as offshore/subsea pipeline. Subsea pipeline is the primary means of evacuation of oil and gas produced in offshore field from offshore to onshore. Subsea pipelines are categorized as below for ease of identification of different sections of pipeline network in offshore.

- **Infield pipelines:** Subsea pipeline meant for transporting the produced liquid containing oil, gas, and water (multiphase flow) from well to pipeline manifold to process platform. In some cases, produced liquid is directly transported from well to process platform. These lines are also called flowlines and feeder lines. Infield lines are also used for transporting treated water or gas for secondary recovery system by water injection or gas injection system, respectively.
- **Export pipelines:** Pipelines used for evacuation of processed oil and gas either together or separately from offshore to onshore terminal for further activities. These pipelines may have single phase (only gas or oil) or multiphase flow (mixture of oil and gas). In real scenario, most pipelines have multiphase flow.
- **Transmission/cross country pipelines:** Pipeline carrying oil and gas from one installation to another on commercial basis.

4.1.1.5 Processing Platforms

A typical oil platform process flow includes:

- Incoming or outgoing pipeline isolation system (like subsea isolation valve, SSIV)
- Separation equipment
- Process heating
- Process cooling
- Crude dehydration
- Produced water treating systems
- Gas handling compression
- Gas dehydration
- Safety and fire detection and fighting

FIGURE 4.3 A typical process platform in the Gulf of Mexico [2].

- The basic function of process equipment and system on any installation is to separate the produced liquid in oil/condensate, gas, and water and stabilize for onward transportation to onshore for further processing. Custody transfer/measurement system forms a part of process equipment and system on any installation. Custody transfer normally occurs at the point of delivery from installation and at the receipt point of installation.
- There are no major differences between the process equipment (oil and gas separators, freewater knockouts, gas scrubbers, pumps, compressors, etc.) installed on an offshore process platform and onshore installation. Preference is given to compact and lightweight equipment and component with enhanced or better corrosion resistance properties (Figure 4.3).

4.1.1.6 Export Pipelines/Tankers for Evacuation of Oil and Gas

Offshore pipelines are more expensive and difficult to build than onshore pipelines, and if the oil and gas field is small, it may be uneconomical to use them. In some areas, long distances or unstable seafloor conditions may make it impractical or impossible to lay pipelines. Instead, tankers can be used to transport oil to shore. If liquefied using special processing equipment, natural gas can also be shipped in tankers from one port to another. Special re-gasification facilities at the receiving port return the liquid to its gas form for cross-country shipment by pipeline. New sources and rising demand for oil and gas during the last half of the 20th century meant shipping larger quantities for longer hauls. To make long-distance transportation more cost-effective, producers also wanted to use the largest carriers the ports could manage. Ultimately, tanker manufacturers developed super tankers or "very large crude carriers (VLCCs)" that measure up to four football fields in length (Figure 4.4).

FIGURE 4.4 A very large crude carrier tankers [3].

4.1.2 Maintenance and Supply

Typical oil and gas processing platform is required to be self-sufficient in terms of energy, water, accommodation, processing, and product stabilization system. Produced oil and gas is then transported to onshore either by pipeline or tanker. Nowadays, some platforms are installed with electrical power fed from shore through subsea electrical transmission as it is overall cost-effective.

Wellhead, production manifold, production separator, glycol process to dry gas, gas compressors, water injection pumps, oil/gas export metering, and main oil pumps are the main elements or systems of an oil and gas production process/installation [4]. All production facilities are designed to have minimal environmental impact. Normally, offshore manned platforms are supported by emergency support vessels (ESVs) to provide immediate support during any exigency like man overboard or emergency evacuation etc. Platform supply vessels (PSVs) cater to all the major resourcing and support (Figure 4.5).

The illustrations below show the relationship between oil companies or operators (dark blue) and the different service and supply segments (light blue). The industry also consists of other services (orange), however, petroleum-induced activity in these segments are not considered a part of the service and supply industry.

4.1.3 Essential Personnel/Workforce

Normally, adequate workforce is maintained on every platform, however, multitasking is also common. Some quite common trade designations used globally are:

- Offshore installation manager (OIM) – Final authority on platform for essential decisions regarding operation of platform, including safety of platform.

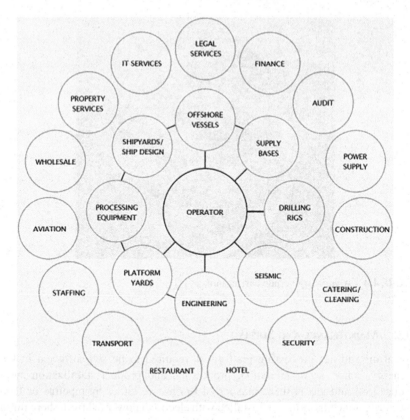

FIGURE 4.5 Maintenance and supply chain directly and indirectly linked with the offshore industry [5].

- Operations team leader (OTL) – Responsible for operations and assists the OIM on all issues related to operation. OTL is assisted by a technical team of control room and other production crew.
- Offshore operations engineer (OOE)/Offshore maintenance engineer – Responsible for all maintenance jobs of platform in consultation with OIM. OOE is supported by multidisciplinary team.
- Safety head/Engineer – Responsible for all aspects of health, safety, and environment. On most installations, paramedics assists them in issues related to health apart from other team members.
- PSTL or Operations coordinator for managing crew changes [4].
- Dynamic positioning operator for navigation, ship, or vessel maneuvering (MODU), station keeping – Available on deepwater floating installation.
- 2nd mate meets manning requirements of flag state operates fast rescue craft, cargo operations, fire team leader – Available on floating installation mainly.

Offshore Production and Transportation

- 3rd mate meets manning requirements of flag state, operates fast rescue craft, cargo operations, fire team leader – Available on floating installation mainly.
- Crane operators – To assist the installation in material transfer on the deck or to and from boat. They are responsible for crane maintenance also.
- Catering crew will include people tasked with performing essential functions such as cooking, laundry, and cleaning the accommodation.

The size and composition of the crew of an offshore installation will vary greatly from platform to platform. Because of the cost-intensive nature of operating an offshore platform and the nature of operations itself, it is important to maximize productivity by ensuring work continues 24 hours a day. This means that there are essentially two complete crews on board at a time, one for day shift and the other for the night shift. Crews also change at regular intervals as per the national law or the requirement of state.

4.1.4 RISKS

Risk is the potential of gaining or losing something of value. Values (such as physical health, social status, emotional wellbeing, or financial wealth) can be gained or lost when taking risk resulting from a given action or inaction, foreseen or unforeseen (planned or unplanned). Risk can also be defined as the intentional interaction with uncertainty. Uncertainty is a potential, unpredictable, and uncontrollable outcome; risk is a consequence of action taken despite uncertainty. To mitigate or minimize risk or threat to operation, normally risk assessment and hazard analysis is done before the start of any operation.

Normally, all external security threat is handled by state authorities as per their risk perception (Table 4.1).

Shutdown Panel
Shutdown panel is a very crucial part of the production operation system which activates and shutdowns the entire production system at the time of any accident or emergency situation. Overall protection of installation is designed with pneumatic shutdown panel as nucleus. Three levels of protection for personnel, production wells, and surface facilities are envisaged.

4.1.4.1 Basic Protection Concepts

Accidental release/leak of hydrocarbon poses a major threat to any production system and installation. Risk analysis for complete installation and system shall be carried out at the design stage and before the start of operations to mitigate such risks.

4.1.5 PREVENTION

Normally, every installation/plant/system is built with two layers of protection, primary and secondary, to mitigate any problems due to accidental release of hydrocarbon.

TABLE 4.1
Risks Associated with Offshore Production Operations (Part A)

Type of Event	Explanation
Anchor failure	Problems with anchor/anchor lines, mooring devices, winching equipment, or fairleads (e.g., anchor dragging. breaking of mooring lines, loss of anchor(s). winch failures)
Blowout	An uncontrolled flow of gas, oil, or other fluids from the reservoir, that is, loss of barrier (i.e., hydrostatic head) or leak and loss of barrier (i.e., BOP/DHSV)
Capsize	Loss of stability resulting in overturn, capsizing, or toppling of unit
Collision	Accidental contact between offshore unit and/or passing marine vessel when at least one of them is propelled or is under tow. Examples include: tanker, cargo ship, and fishing vessel. Also included are collisions with bridges, quays, etc., and vessels engaged in oil and gas activity on other platforms than the affected platform, and between two offshore installations (to be coded as CN only when intended for close location)
Contact	Collisions/accidental contacts between vessels engaged in the oil and gas activity on the platform affected, for example, support/supply/stand-by vessels, tugs or helicopters, and offshore installations (floating or fixed). Collisions between two offshore installations are also included only when these are intended for close location. Contact is result of coming near of two floating or one floating and one fixed offshore installation without any damage or loss of anything
Crane accident	Any event caused by or involving cranes, derrick, and draw-works, or any other lifting equipment
Explosion	Explosion
Falling load	Falling load/dropped objects from crane, drill derrick, or any other lifting equipment or platform. Crane fall and lifeboats accidentally to sea and man overboard are also included
Fire	Fire
Foundering	Loss of buoyancy or unit sinking
Grounding	Floating installation in contact with the sea bottom
Helicopter accident	Accident with helicopter either on helideck or in contact with the installation
Leakage	Leakage of water into the unit or filling of shaft or other compartments causing potential loss of buoyancy or stability problems
List	Uncontrolled inclination of unit
Machinery failure	Propulsion or thruster machinery failure (including control)
Off position	Unit unintentionally out of its expected position or drifting out of control.
Spill/release	"Loss of containment". Release of fluid or gas to the surroundings from unit's own equipment/vessels/tanks causing (potential) pollution and/or risk of explosion and/or fire
Structural damage	Breakage or fatigue failures (mostly failures caused by weather, but not necessarily) of structural support and direct structural failures. "Punch through" also included
Towing accident	Towline failure or breakage
Well problem	Accidental problem with the well, that is, loss of one barrier (hydrostatic head) or other downhole problems
Other	Events other than those specified above

Offshore Production and Transportation

4.1.6 SHUT IN

In the event of loss of containment due to release or leak of hydrocarbon, it is essential to stop feeding the hydrocarbon to the area of release to minimize or eliminate any chance of contamination of environment/fire/explosion. Protective shut-in is always incorporated in design itself to achieve above.

4.1.6.1 Fire and Gas Leakage Protection System

Protection system is designed for automatic detection of any gas leakage before it forms a combustible mixture, with subsequent initiation of shutdown action. Such shutdown includes all ignition sources. This protection is designed in such a manner that it does not allow restarting of units till hazardous condition is removed. In case of fire incidents, detection is followed by automatic initiation of suppression action.

4.1.6.2 Technology Development: HIPS

HIPS: High integrity protection system

HIPS is a high availability fail-safe safety instrumented system designed to achieve a predefined risk reduction, as defined by the Safety Integrity Level (SIL-3). HIPS can replace flare/mechanical pressure relief or mechanical thickness of piping to remove the source of overpressure.

The risk analysis report addresses the following major safety elements:

- Identifying the hazards mitigation and protection provided by the HIPS and full flare system.
- Determining the potential frequency and consequences of each identified hazard.
- Determining the system availability of each of the two overpressure protection systems.
- Quantifying the associated risks for both overpressure protection solutions in terms of IRPA and PLL (Figure 4.6).

4.1.6.3 Surface Facility Protection

A safety analysis or hazardous operability (HAZOP) analysis of surface facilities including all systems and equipment on board is carried out. All possible hazards and interrelation between various parameters are identified and listed. The functional chart thus evolved is the safe safety analysis and function evaluation [4].

A hazard and operability study (HAZOP) is a structured and systematic examination of a complex planned or existing process or operation to identify and evaluate problems that may represent risks to personnel, system, or equipment. Figure 4.7 explains the procedure of HAZOP for blast hazard/fire hazard. The intention of performing a HAZOP is to review the design to pick up design and engineering issues that may otherwise not have been found or may result due to integration of different systems. The technique is based on breaking the overall complex design of the process into a number of simpler sections called "nodes", which are then individually reviewed.

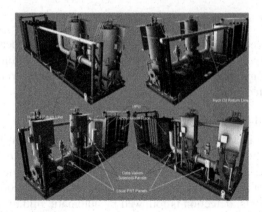

FIGURE 4.6 Three-dimensional model of HIPS system [6].

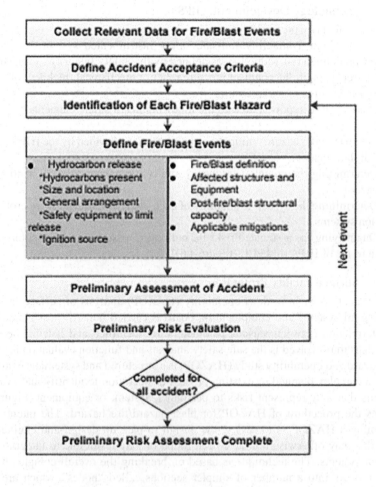

FIGURE 4.7 HAZOP tree construction for fire/blast hazard [7].

WELL CONTROL-SCHEMATIC

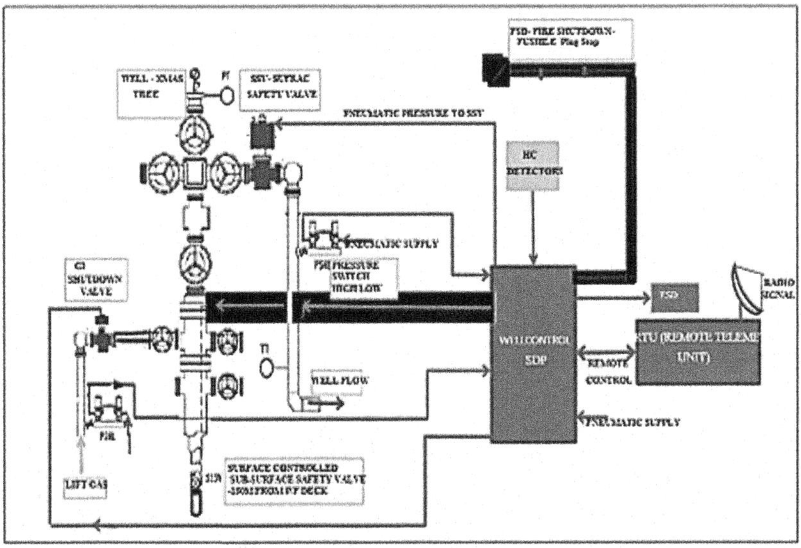

FIGURE 4.8 A well control process with help of shutdown panel [7].

4.1.6.4 Well Control and Protection

Wells need to be closed quickly to prevent spillage and sea pollution in the event of sudden emergency at the production platform. Wellhead shutdown panel is designed for well-defined sequential operation of various valves installed on wellhead including surface and subsurface safety valve for control of flowing well. Figure 4.8 explains the automated well control shutdown panel.

4.1.7 SCADA – An Essential Part of Digital Oil and Gas Field

Supervisory Control and Data Acquisition System (SCADA) has revolutionized the complete process controlled industry and has become an integral part of offshore operations as well. This is the most common intelligent system utilized to maintain and optimize the production at offshore and as well as onshore platforms. Figure 4.9 explains the pyramidal steps of SCADA utilized to optimize the production, while Figure 4.10 explains the process of data acquisition in SCADA system.

Flow chart for Decision Making based on data acquisition

The detailed process level of SCADA system is explained in the below sections.

4.1.7.1 Process Levels in SCADA

4.1.7.1.1 Manual Process Level

This is the initial reference point which assumes a fully manual process. The data are not being acquired in real time. Basic surface sensors (pressure (P), temperature (T), and volume (V)) are used, and readings are recorded and annotated by people when

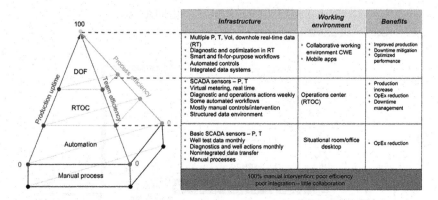

FIGURE 4.9 A digital oil field pyramid explaining the entire process [8].

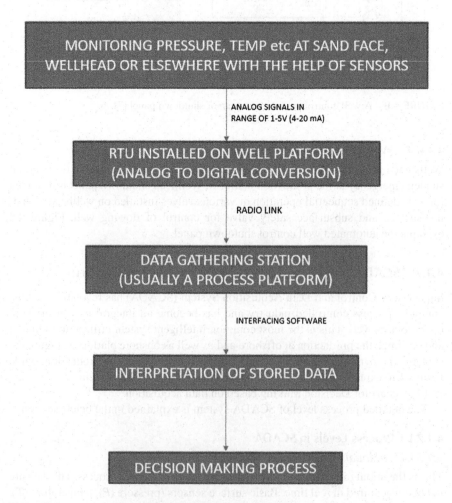

FIGURE 4.10 An entire cycle of data acquisition cycle [7].

Offshore Production and Transportation

they visit the operation site. Production well tests are conducted periodically on demand. The data are exchanged between different disciplines by email or in a repository of shared folders. The engineering workflows are performed manually by each discipline (in silos, no integration). Monitoring is performed monthly, and diagnostic and optimization are performed randomly (2–3 times a year). Communication is by phone, email, and meetings. Asset team collaboration is low.

4.1.7.1.2 Automation Level

The real-time data is gathered only from basic surface sensors (P, T), and the information is gathered using wireless technology. Production well tests are taken monthly or on demand. Data are centralized in a SCADA/historian storage center using industry-standard protocols. Some engineering workflows are automated. Monitoring is performed daily, while diagnostic and optimization processes can be performed monthly. Team discussions of production issues are conducted in meetings and situation rooms, and collaboration starts to improve [9].

4.1.7.1.3 RTOC Level

Most surface locations in a field have real-time sensors, including flow meters. For wells without flow meters, virtual metering is used for the entire field that can provide production data for individual wells. Data are sent using wireless or WiMAX technologies (to support high-volume data traffic). The data are centralized in the SCADA/historian storage center using the industry standard protocols. Most engineering workflows are automated with advanced algorithms to provide alarms and alerts. Monitoring is performed in real time, while diagnostic and optimization can be performed weekly or monthly [9]. The operation includes a dedicated real-time operations center, with a dedicated staff. Collaboration is significant but not optimal. Communication with field operations staff is via cell phone and by texting.

4.1.7.1.4 DOF Level

The operations is exactly the same as an RTOC; however, most engineering workflows are intelligent and with predictive capability to generate advice and guidance. Monitoring is performed in real time with exception-based surveillance, while diagnostic and optimization can be performed daily with an advisory system to prevent production downtime [10]. There is a dedicated collaboration working environment (CWE) with dedicated staff and complete workflow mobile communication with field operations staff. Collaboration reaches very high levels with synergy between disciplines. Communication with field operations staff is via closed-circuit TV, video, and chatting.

4.1.7.2 Instrumentation, Remote Sensing, and Telemetry of Real-Time Processes

This area focuses on the equipment and technology in the physical oil and gas operations, both on the surface and downhole, required for telemetry, remote collection, and transmission of data required to monitor, optimize, and automate operations. The wellhead includes a series of mechanical or electronic devices (gauges) to measure real-time pressure, temperature, fluids, and other special data such as chemicals, solids detection, and radiation (Figure 1.14). Downhole locations are equipped with another family of sensors especially designed to work in high-temperature and high-pressure conditions. Sensors are connected to electrical cables that send analog pulses to a control panel located close to the wellhead. The control panel consists of many hardware components for the analog to digital signal conversion [11]. A key component includes remote terminal units (RTU) and programmable logic controllers (PLC), which perform similar functions. They are connected to sensors with cables and send digital data to the transmission hardware using wireless equipment that includes ethernet, switchboards, WiMAX (microwave signals), and routers all connected to a CPU, often powered by a solar panel.

Figure 4.11 explains an offshore wellhead with a series of mechanical or electronic devices to measure real-time pressure, temperature, fluids, and other special data such as chemicals, solids detection, and radiation. Furthermore, Figure 4.12 explains the working of SCADA system along with its wireless features.

4.1.7.2.1 Data Management and Data Transmission

Located in the SCADA terminal, real-time signals from the field are gathered by cellular modems and sent to a family of servers. The servers use multiplex software to organize and store the data in different structured layers under a series of information technology (IT) industry protocols. The software that does this data collection and aggregation is referred to as a historian, which accumulates time data, Boolean events, and alarms in a database, which can be used for many visualization solutions. The data are previously QA/QCed, cleaned, and conditioned using a series of

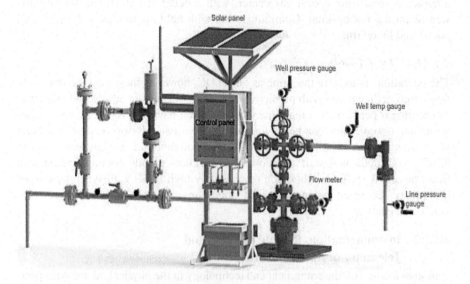

FIGURE 4.11 A wellhead with a series of mechanical or electronic devices (gauges) to measure real-time pressure, temperature, fluids, and other special data such as chemicals, solids detection, and radiation [12].

Offshore Production and Transportation

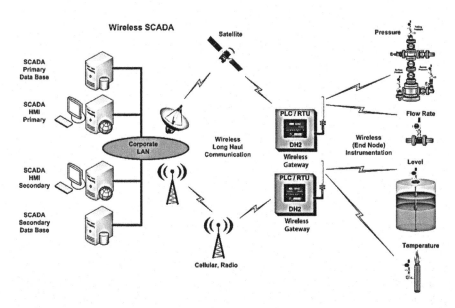

FIGURE 4.12 The working of SCADA system and wireless [12].

algorithms (data reduction, wavelet filtration, and missing data interpolation) that filter the data from signal abnormalities such as noise, spikes, outliers, and frozen data. The historian commonly feeds a repository or master database, such as the Structured Query Language (SQL) or Oracle. Other types of data, such as mechanical equipment, interventions, tubing scans, and gyro scans, are unstructured and stored in well files are well databases.

4.1.7.2.2 Workflow Automation

Traditionally, geoscientists and various engineering disciplines (production, reservoir, facilities, etc.) spent considerable time gathering data from disparate sources for input into their mostly manual workflows.

Engineers generally use models developed in commercial software applications to reproduce the oil production process. However, even these software models required complex manual workflows that consumed engineer's time, for example, collecting data from different sources (spreadsheet, text, tables, figures, historian, etc.); filtering data from noise; performing repetitive, error-prone tasks to update models (e.g., manual data entry); reconciling the data and calibrating the model; and running different model scenarios.

Workflow automation uses high-level programming language routines to connect these manual processes, so that models can be automatically populated and updated. Automation is just part of the DOF requirement for workflow construction.

DOF solutions also require that engineering workflows are intelligent enough to capture in real time alarms and alerts to generate prompt actions, update engineering applications, and deliver right-time monitoring, diagnostics, and process optimization that deliver operations guidance at the field level.

4.1.7.2.3 Well Automation

Generally, in offshore, a number of control systems are employed for automation and effective management of oilfield operations. Out of these systems, SCADA is a very comprehensive and sophisticated automation system. SCADA system is utilized for all kinds of production system installed with different artificial lift systems. In offshore production system, generally installation of gas lift system is very common due to abundant availability of gas and lack of storage and transportation facilities for gas. The lack of gas storage and transportation facility is due to its expensive nature and constraint of space and isolation from the land. Thus, to explain the well automation for production optimization has been explained for a gas lift systems.

4.1.8 AUTOMATED GAS LIFT OPTIMIZATION IN OFFSHORE

Gas lift automation is a key to get rid of daily struggle of gathering enough data to analyze field production. Combination of automation equipment and computer software allows operators to control each well's lift gas injection rate. It enables the field engineer to monitor the well in real-time from the computer. Continuous gas lift automation along with wellhead surveillance system installed on offshore platform to ensure technocommercial profitability in oil production.

Gas lift optimization in offshore is usually implemented through continuous gas lift method. However, major problem encountered with continuous gas lift is maintaining an optimum gas injection rate assigned to each well [13]. This is attributed to the fact that, as injection gas is limited in offshore facilities, optimum gas rate cannot be provided to each well. In past, optimum gas rate is considered the rate at which the well would yield maximum production. Nowadays, optimum gas injection rate is recognized from gas lift performance curve where the cost of additional injection gas exceeds the anticipated profit that would be made via increased oil production. Figure 4.13 explains an overview of well automation and Figure 4.14 displays the gas lift automated system for offshore production operations.

The primary objective of gas lift optimization system is to inject less gas to less productive system but continue to inject the optimum rate to most productive wells in case of limited gas supply.

The optimization system has four main tools that work together to provide overall benefit.

1. Constant gas lift injection rate

   In the automation system, the system and software assisted tools constantly measures gas lift rates and adjusts injection choke according to set point. Maintaining a constant injection rate decreases amount of slugging in the wells, and thus reduces production process instabilities.
2. Real-time wellhead surveillance

   This tool enables engineers to monitor temperature and pressure data of individual wells.

Offshore Production and Transportation

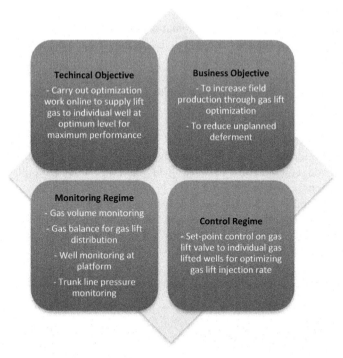

FIGURE 4.13 An overview of well automation [13].

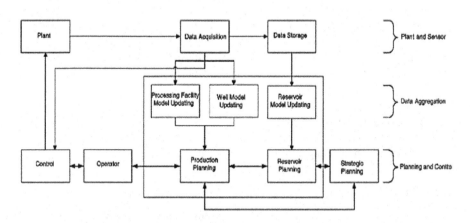

FIGURE 4.14 Flowchart showing typical offshore gas lift optimization network [13].

3. Optimization well testing

It is imperative to run optimization well test on each well to determine optimum gas lift injection rate on each well. The system and software assisted tools then determine the optimum injection rate by analyzing the fluid flow versus gas lift rate data.

4. Data to desktop

This tool transmits all the automation data back to office and stores it in a database. The engineer can access the data and customize the interface to display specific well data.

Thus, production process automation yields the following benefits of automated SCADA system:

1. Gains in oil production.
2. As operator can access the same data and system on desktop, there is improved collaboration between field planning, well surveillance, and operations.
3. Capture and save the history of lifecycle of each well.
4. Wellhead and well-test data is online and easy to access.
5. Availability of historical data to perform analysis, modeling, and optimization.

Now after understanding the offshore production operations, we will discuss the next step of processing the produced fluids.

4.2   PROCESSING IN OFFSHORE

In general, three phase fluids are encountered from offshore wells (oil, water, and gas). The produced well fluids are oil, water, and/or gas (if present), all of which must be separated and individually treated before further dispatch or disposal. The well fluids are collected at the topside from the wells using a complex network of subsea pipelines and risers and treated so that they can be further dispatched to the required destination for commercial usage. For processing these fluids large process platforms; these platforms generally contain the following process modules (Figure 4.15):

1. Fluid separation and oil dispatch
2. Gas compression and dehydration
3. Produced water treatment plant
4. Seawater conditioning and injection system

4.2.1   OIL TREATMENT

The target of the separation of reservoir fluids from undersea wells is to produce a gas stream free of C3+ hydrocarbons to the maximum extent and a crude oil stable at storage conditions. Indeed, the produced crude oil must not vaporize when delivered into the storage tank and in the event of small variations of storage pressure and/or temperature.

Aims are often accomplished by progressively reducing pressure and temperature of the fluid through a multistage separation consisting of a sequence of two or three separators, which forms a configuration termed separation train. The topside contains a series of machines and equipment called a train. Produced fluids are treated through this train. Each train consists of a production manifold, well

Offshore Production and Transportation

FIGURE 4.15 An integrated process platform with living accommodation [3].

fluid heater, inlet separator, crude oil manifold, crude oil heater, surge tanks, and oil pumps [14]. The treatment occurs in the following manner:

1. Well fluids arrive at the production manifold through subsea pipelines and risers.
2. A de-emulsifier is dozed into the production manifold, which will improve the oil–water separation and break any emulsions formed between the fluids.
3. The dozed well fluids are passed through the well fluid heater, which provides heat and further improves separation. (Hot oil is circulated in the shell side, while well fluids flow through the tubes.)
4. The hot well fluids are sent to the inlet separator. The inlet separator separates the fluids with aid of gravity, and improved separation is achieved with chemicals and heat.
5. Separated gas is taken out through the top and is sent for compression and dehydration.
6. Separated water is siphoned through the bottom and sent for conditioning. Here we are performing partial separation, and pump oil, water, and gas mixture to land using special pumps called main oil line (MOL) pumps.
7. The oil from the inlet separator is further heated with a crude oil heater and sent to the surge tank with an oil manifold.
8. Surge tank is maintained at a lower pressure to stabilize crude, that is, to remove maximum associated gases from crude oil.
9. Oil from surge tanks can be either pumped directly with MOL pumps or can be diverted to third-stage separators (Surge tank 3).

10. Separated crude oil is pumped with CTP/MOL pumps to export trunk lines.
11. Separated gas is diverted to gas compression module after boosting the pressure LP booster compressor [13].
12. Excess gas after internal consumption is delivered to export gas pipeline after compression and dehydration on platform for further processing on land.
13. Separated water is diverted to produced water conditioning unit.
14. The pumps dispatch the oil from the platform using pipelines connected to the land, possibly a port or a refinery (if one is situated sufficiently close).

4.2.1.1 Important Notes

The surge tank is placed at lower pressures than the separator to stabilize the crude oil, that is, remove as maximum associated gas from the oil.

There can be more stages of separation as per requirements, otherwise oil is usually sent to the pumps for dispatch (Figure 4.16).

In the early stages of the platform, when pipelines have not been laid, oil is dispatched by tankers.

4.2.1.2 Loading of Tankers

Tankers are filled by using a set-up of submarine pipeline, flexible hose, and a floating tank/buoy. The buoy is held at a fixed location by anchors and moors the tanker to it with a mooring rope.

4.2.2 Gas Treatment

Gas previously separated from oil is taken for compression and dehydration prior to usage or dispatch. The separated gas can be used for either one, two, or all three utilities of export, lift gas, and internal usage.

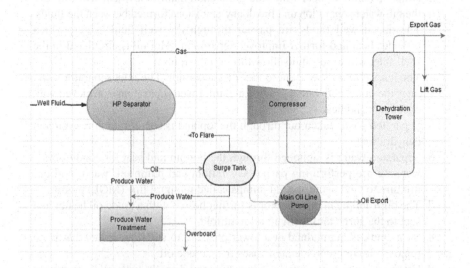

FIGURE 4.16 Oil and gas processing on offshore platform.

Dehydration is the most important treatment for the produced gas as further processes may cause formation of gas hydrates. Gas hydrates are formed when moisture-laden gases are subjected to high pressures and low temperatures. If formed, gas hydrates can solidify in areas that can hamper smooth oil field operations such as fluid transportation in pipelines and compressors. By dehydrating the gas, these potential problems are eliminated. Removal of water is done for the following reasons:

- Natural gas can combine with liquid or free water to form solid gas hydrates that can plug pipelines or valve fittings.
- Water can condense in the pipeline causing corrosion.
- Water vapor increases the volume and decreases the heating value of the gas.
- The maximum water content is 7 lb H_2O/MMscf. There are several methods of dehydrating natural gas. The most common of these are liquid desiccant (glycol) dehydration, solid desiccant dehydration, and refrigeration (i.e., cooling the gas) (Figure 4.17).

The glycol dehydration process is widely used in offshore platforms for gas treatment (Figure 4.18).

4.2.2.1 Gas Dehydration Follows the Steps Below

1. Wet feed gas is entered a tray tower; the gas rises from the bottom while lean (water-free) glycol is sprayed from the top.
2. The glycol mixes with the gas and dehydrates the gas, allowing dry gas to vent through the top while water-laden rich glycol is collected into the bottom of the tray tower.

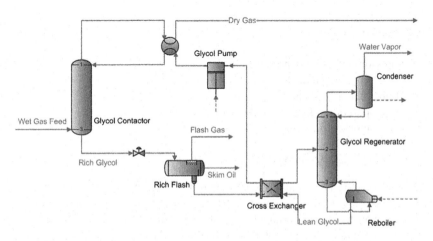

FIGURE 4.17 Glycol dehydration unit [15].

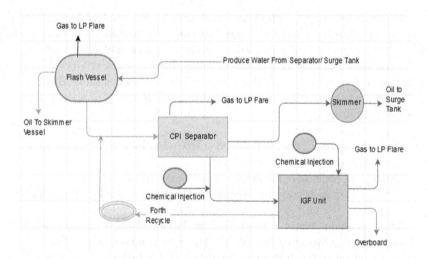

FIGURE 4.18 Produced water treatment unit.

3. The cool, rich glycol is sent to a flash tank to remove any gases in the glycol. Any flash gas obtained is sent for flaring and any low pressure gas is compressed and mixed in the gas stream.
4. The cool glycol is sent through a reflux coil for heating, and then passed through a rich glycol-lean glycol heat exchanger, which further heats the rich glycol.
5. When rich glycol is re-introduced into the second tray tower, it is hot enough to allow the water to escape as steam, which is vented. The glycol, now stripped of water, falls to the bottom.
6. This lean glycol is sent for re-boiling and removal of flue gases.
7. Once the flue gases are removed, the lean glycol is sent through the heat exchanger for cooling and then re-supplied into the first tower.

Triethylene glycol (TEG) is the most commonly used glycol for such an operation. The benefits of TEG include:

- Superior dew point depression
- Easy to regenerate up to 99% of utilized chemicals
- Higher decomposition temperature
- High operation reliability and low operating costs
- Low vaporization losses

4.2.3 Produced Water Treatment

The water recovered through the separation train and gas processing contains traces of oil and other impurities. Produced water stream can be reused through re-injection in the reservoir to enhance production, or it is discharged in the sea after treatment. In the latter event, impurities and oil must be removed by several processing steps. In particular, oil concentration must be lowered below 40 mg/L.

Offshore Production and Transportation 121

Water discharge standard in offshore and onshore are different. This is because in offshore very less land is available compared to onshore. To achieve the standard of onshore in offshore it required huge amount of capital investment that will adverse impact on ecumenical aspect of the project.

The most undesired component of well fluids is reservoir water. However, disposal of this fluid is not easy. Produced water is significantly different from water present on the surface. Further, contact with oil makes treatment even more necessary before it can be disposed into sea.

International regulations of quality must be followed, and the water must be treated to acceptable levels before being discharged into the sea.

The water conditioning/treatment unit consists of four units, namely:

1. Flash vessel

   This vessel receives water from the inlet separators and surge tanks, and is maintained at 0.8 kg pressure. In the flash vessel, most dissolved gases are flashed out, which are then routed to the LP flare header. Associated oil from the produced water is sent to a closed drain and the oil is collected in the sump caisson.

2. CPI separators

   The CPI separators (corrugated plate interceptor) collect oil from the water sent from the flash vessel. These are plates installed in parallel, and the oil collected from these CPI separators is collected in a tank. This collected oil is later pumped to the oil manifold; gas is flared off.

3. IGF unit

   Water from the CPI separators flow into the IGF unit. The IGF unit is a tank that uses motor-driven agitators to aerate gas bubbles in the water. These bubbles float the oil droplets to the surface, which is then collected and pumped to the CPI separators [13].

4. Sump caisson

   This vessel collects water sent from the IGF unit. The vessel has an open bottom from which water continuously drains into the sea. Oil floating in the surface of the sump caisson is collected in the blow caisson. The oil is then lifted and flown into a skimmer by gas injection.

4.3 SEA WATER INJECTION

Sea water is often used in offshore platforms when the necessity of water injection arises. Water injection is done to maintain reservoir pressures as well as during water flooding. To prevent damage to the reservoir, strict quality standards are imposed on water quality. Additional chemicals are also added to protect pipelines that carry the water to the wells and well platform.

The main components of water injection systems are:

a. Sea water lift pumps
b. Coarse filters
c. Fine filters

122   Offshore Operations and Engineering

d. De-oxygenation towers
e. Booster pumps
f. Main injection pumps
g. Chemical dosing system

The standard chemicals used during sea water injections are:

a. Flocculants
b. Scale inhibitors
c. Corrosion inhibitors
d. Chlorination
e. Bactericide
f. Oxygen scavengers

The water injection system operates according to the following procedure:

1. Sea water is lifted with sea water lift pumps and fed to the filters for filtering.
2. Coarse filters are able to filter particles up to $20\,\mu m$, while the fine filters filter particles up to $2\,\mu m$. Flocculants and coagulants are added to promote coagulation and filtration of suspended particles.
3. Filtered water flows to de-oxygenating towers for oxygen removal. Deoxygenating prevents formation of aerobic bacteria colonies in the injection flowlines. Vacuum pumps and oxygen scavenging chemicals are used to de-oxygenate water.
4. Booster pumps take de-oxygenated water and feed it to the main injection pumps.
5. Other chemicals like scale and corrosion inhibitors are dozed into the water during booster pump discharge.
6. The main injection pumps send the treated water to subsea pipelines and into water injection wells [16].

The next step after processing is the storage of process fluids for dispatch. In the following section, offshore storage system is explained in detail.

4.4 OFFSHORE STORAGE

Offshore storage is normally required because there is inevitably a question of down-time (time during which a system is not operational) associated with offshore loading concepts. The normal method of crude evacuation is by shuttle tankers which may be loaded directly from a loading system or via a storage vessel. If no storage is provided and adverse weather prevents shuttle tanker loading, the platform supervisory personnel have no option but to shutdown field production [9]. The field reservoir characteristics are not always consistent with this stop-start type of production, so some element of buffer storage must be considered. Several aspects of crude oil evacuation must be examined before a suitable buffer storage system is selected. Among the factors to be considered are:

- Storm occurrence interval and persistency
- Oil throughput
- Distance of the field from port of discharge
- Speed of the shuttle tanker
- Number and capacity of the shuttle tanker(s)
- Efficiency of discharge port equipment
- Loading system maintenance downtime (hoses, hawser, etc.)

The environmental factors will of course have a bearing on the type of storage structure selected, but the factors listed above govern the quantity of storage required.

Four basic structures suitable for offshore storage include:

- Tankers
- Barges
- Articulated column
- Spar

4.5 TRANSPORTATION OF OIL AND GAS

Pipeline transportation of oil and gas remains the first choice even in offshore. However, in case of marginal offshore field (where pipeline may not be cost-effective) or during extended testing phase when pipeline is not available, oil can be stored in a floating tank and transferred to shuttle tanker through buoy for processing on land (Figure 4.19).

4.5.1 OIL TANKERS

Crude oil tankers are large ships used to transport crude oil from the Middle East, Africa, and Latin America to refineries worldwide. Tankers that carry refined products are known as product tankers and carry refined petroleum products from refineries to distribution locations.

FIGURE 4.19 Testing/marginal field [17].

FIGURE 4.20 A commercial oil tanker. It is important to note that oil tankers are very large, with small ones being around 200 m in length and large ones up to over four 400 m in length [17].

Oil tankers can vary in size, although smaller vessels are generally used to transport refined petroleum products, whereas much larger tankers are used to transport crude oil. These larger ships tend to be used more for crude oil as they reduce the cost per barrel of oil transport. Tanker sizes are expressed in terms of cargo tones, or how much cargo they can carry. Larger crude oil tankers carry significantly more and are generally utilized in international crude oil trade [18] (Figure 4.20).

4.5.2 Pipelines

Pipelines are systems of large transportation pipes generally below ground on the sea bed that serves the purpose of transporting oil and natural gas within extensive distribution networks. These lines vary in diameter depending on their use and are generally located underground. In Canada there is an estimated 825,000 km of lines that serve to transport natural gas, liquefied natural gas products, crude oil, and other refined petroleum products [18].

4.5.3 Floating Production, Storage, and Offloading (FPSO)

Small, marginal, and isolated fields, which do not justify the high cost of platform and pipeline, are produced through an integrated system called FPSO.

This type of production support consists of a tanker converted for production operation with a permanent yoke arrangement. The tanker is allowed to weathervane around the SPM by means of a fluid swivel arrangement at the yoke/SPM interface. A converted tanker is used because it provides the cheapest form of floating production and already has existing oil storage capacity. Because these structures have appreciable motions, the wells are typically subsea completed and connected to the floating unit with flexible risers that are of either composite material or rigid steel with flexible configuration. While the production unit can be provided with a drilling unit, typically the wells are predrilled and the production unit brought in to

FIGURE 4.21 Floating production, storage, and offloading (FPSO) with external turret [19].

carry only a workover drilling system. Previously, FPSOs in shallow waters and in mild environment had spread mooring systems. As more FPSOs were designed and constructed or converted (from a tanker) for deepwater and harsh environments, new and more effective mooring systems were developed including internal and external turrets. Some turrets were also designed to be dis-connectable so that the FPSO could be moved to a protective environment in the event of a hurricane or typhoon (Figure 4.21).

Advantages of FPSO include:

- FPSOs have integral oil storage capability inside their hull. This avoids a long and expensive pipeline to shore.
- Large area for process equipment.
- Can explore in remote and deepwater as well as in marginal wells, where building fixed platform and piping is technically and economically not feasible.
- Easy loading of shuttle tanker from FPSO.
- Easily converted to production support.

4.5.3.1 Single Buoy Mooring

Single buoy mooring (SBM) (also known as single point mooring or SPM) is a loading buoy anchored offshore that serves as a mooring point and interconnect for tanker loading or offloading liquid products. They are capable of handling any size ship, even very large crude carriers (VLCC), where no alternative facility is available.

In shallow water, SPMs are used to load and unload crude oil and refined products from onshore and offshore oilfields or refineries usually through some form of storage system. These buoys are usually suitable for use by all types of oil tanker. In deepwater oil fields, SPMs are usually used to load crude oil direct from the production platforms, where there are economic reasons not to run a pipeline to the shore [19].

REFERENCES

1. "Oilfield review," *1989*. schlumberger oilfield review.
2. *processplatform/gulfofmexico*. Available at www.stateoil.com/processplatform/gulfofmexico.com.
3. *offshore-oil-amp-gas-production*. Available at http://docshare.tips/offshore-oil-amp-gas-production-_5767ae19b6d87fa6918b4582.html.
4. *Oil platform*. Available at https://en.wikipedia.org/wiki/Oil_platform.
5. *"Exxon Mobil Offshore Operations Manual,"* 2016.
6. *"World Offshore Accidental Database,"* 2000.
7. *Saudi Aramco Development Program*. Available at www.saudiaramco.com/en/careers/saudi-applicants/our-offer/employee-training.
8. *A digital oil field Pyramid explaining the entire process*. Available at www.petrowiki.org.
9. *What-is-scada*. Available at https://oleumtech.com/what-is-scada.
10. G. Carvajal, M. Maucec, and S. Cullick, Introduction to digital oil and gas field systems, in *Intelligent Digital Oil and Gas Fields*, Elsevier, Gulf Professional Publishing, UK (2018), pp. 1–41, https://doi.org/10.1016/B978-0-12-804642-5.00001-3
11. *Remote-monitoring-and-control*. Available at http://bluetickinc.com/remote-monitoring-and-control/resources/.
12. *Oilfield-monitoring*. Available at http://petrocloud.com/solutions/oilfield-monitoring/.
13. P.W. Abshire, D.J. Hoge, and C.H. Neeley, Programmable controllers integrate oil field SCADA and automation, in *37th Annual Conference on Petroleum and Chemical Industry*, Houston, TX (1990).
14. N. Mitra, *Principles of Artificial Lift*, Allied Publishers Ltd, New Delhi (2012).
15. *Central-processing-platforms*. Available at www.thepiping.com/2015/08/central-processing-platforms-cpp.html.
16. *glycolunit*. Available at www.wikipedia.org/glycolunit.
17. *Transportation_of_oil*. Available at https://energyeducation.ca/encyclopedia/Transportation_of_oil.
18. D.P.V. Li, *Offshore Engineering Handbook* Series.
19. *Starzz Offshore Structure Website*.

5 Utilities and Support System

Although the technology of drilling or production does not change drastically between onshore and offshore environments, drilling or production is entirely different in both these environments. This is because the amenities and support services freely available onshore have to be constructed in offshore operations, which is not just limited to living accommodations.

Offshore installations are self-sufficient in every respect including energy, water, accommodation, etc. Product evacuation is done either through dedicated pipelines or tankers.

All production facilities are designed to have minimal environmental impact. Normally, offshore manned platforms are assisted by emergency support vessels (ESVs) to provide immediate support during any exigency such as man overboard or emergency evacuation. Platform supply vessels (PSVs) cater to all major resourcing and support.

Because of the cost-intensive nature of operating an offshore platform and the nature of operations, it is important to maximize productivity by ensuring that work continues 24 hours a day. This means that there are essentially two shift crews onboard at any time, one for day and the other for night shift. Crews also change at regular intervals to maintain continuous operations.

5.1 LIVING ACCOMMODATION

Accommodation on a drilling or production platform has improved considerably in recent years. Some set-ups can be compared to that of a dormitory with four to eight people often sharing a room with bunk beds and private rooms (for head engineers). Laundry facilities are provided after each shift change to ensure cleanliness. Dormitories create a very close relationship between workers with some describing their roommates as a "second family". All room cleaning services are done by staff people during shift change, ensuring cleanliness for the new shift. Figure 5.1 displays the living accommodation on offshore platforms [1].

Food
Most offshore installations have self-service style canteens/mess with a wide variety of food options, including fresh food such as salad, as well as vegetarian and non-vegetarian food options, which are delivered to the rig via helicopters or boats. Each offshore rig has a dedicated team of kitchen staff who prepare high-quality food around the clock. Special arrangements are also made during festivals.

Various galley equipment for cooking, refrigeration, preparation, and laundry are available on offshore platforms [3].

FIGURE 5.1 Living room [2].

Cooking Equipment	Preparation
Oven	Mixer
Cooker	Peeler
Boiling pan	Slicer
Frying pan	Toaster
Griddle	Cupboard

Figure 5.2 displays the high standard of food available at offshore platforms.

5.1.1 Medical

All employees are required to undergo a medical examination to ensure that their health is optimum for work on an offshore platform. Paramedic and medical personnel are available onboard. Moreover, transport facilities are on standby for quick medical evacuation from offshore facilities, if required.

FIGURE 5.2 Food in offshore installations [4].

Utilities and Support System

5.1.2 SMOKING AND ALCOHOL

Smoking on an offshore platform is only allowed in designated areas. Although ignition sources are not allowed on the installation, matches are supplied at the designated smoking areas. No alcohol and drug policy is the norm on all offshore facilities and falls under prohibited category.

5.1.3 ENTERTAINMENT AND RECREATION

Realizing the stress involved in working offshore, all offshore facilities have adequate recreational facilities as per the company policy to keep the onboard staff entertained when not on shift. For example, small movie theaters, pool tables, game zones, gymnastics, large screen TVs, video game consoles, etc. are provided on offshore facilities.

5.2 POWER GENERATION [5]

Power generation systems provide electrical power for drilling and production, as well as all other platform activities and operations.

5.2.1 FUEL GAS SYSTEM

The fuel gas system comprises suction knockout drums (KODs), compressors, heaters, scrubbers, filters, and superheaters. The power generation system PG compressor and in some cases main oil line (MOL) pumps have gas generators as drivers. Fuel gas supply to these generators at a particular pressure and temperature is vital. Moreover, no liquid should be carried over, which has potential for creating a hot spot, choking the injectors, and increasing the combustion temperature. Every unit has its own fuel gas valve, which requires supply pressure at a predetermined range as fuel gas, otherwise it can stop supply, and the unit will trip. Hence, fuel gas conditioning before it is sent to gas generators is essential.

Fuel gas can be taped from the separators or from the outlet of the glycol towers. If the separation pressure for fuel gas is very low, it needs to be compressed. This is done by a dedicated fuel gas compressor, which is a gas-engine-driven, reciprocating-type compressor. Gas from the separators is scrubbed by the suction KOD prior to compression. Depending on the requirement, the compressor can be bypassed, and on increasing the separator pressure, gas can be directly supplied to the conditioning skid. The line from the tower outlet has a pressure control valve (PCV), in addition to a shutdown valve (SDV) to safe guard equipment from damage, at a much higher pressure. Subsequently, the gas is heated up to 55°C–60°C. This electric heater is equipped with thermal switch high (TSH), which automatically cuts off the heater in case of actuation. Gas from the heater enters the fuel gas scrubber, where the liquid droplets are knocked off. A PCV downstream the heater maintains the pressure in the fuel gas manifold.

As a second line of protection, the conditioning skid is equipped with filters downstream of the PCV to ensure that the gas is absolutely liquid-free. The gas from the filter enters the superheater where the temperature is raised to approximately

70°C–75°C. The fuel gas heater is quite long, even though jacketed, but any temperature drop can result in condensation; therefore, a superheater is used to increase the temperature. Moreover, the superheater is equipped with a temperature switch which cuts off the electrical supply on actuation. The individual units have their own toppings from the header and are equipped with SDV and blowdown valve (BDV). In case the units shut down, the SDV closes and the BDV opens such that the gas downstream of the SDV gets depressurized. The fuel gas header is also equipped with a BDV, which actuates in case of emergency shutdown (ESD) or fire shutdown (FSD), depressurizing the entire system.

5.2.2 Utility/Diesel Generators

Offshore installations/production platforms are provided with utility/diesel generators for power requirement in case of power or ESD including black start-up. These generators supply power to essential systems to restore power supply after outage or black start-up of the platforms, including emergency lighting systems and gas turbine instrument air system. Figure 5.3 displays a diesel generator present on offshore platforms. This is the main power supply on a drilling platform as gas remains unavailable during the drilling phase.

5.2.3 Gas Turbine Generator

Aero-derivative and industrial-type gas turbines from various manufactures, for example, Rolls Royce, GE, Solar, Siemen, Man GHH, etc., are used in offshore facilities, either for power generation or as prime mover for gas compressor/main oil pumps/main water injection pumps. These machines are considered critical for platform and field operations.

The compressed air is mixed with fuel injected through nozzles. The fuel and compressed air can be premixed or the compressed air can be introduced directly into the combustor. The fuel–air mixture ignites under constant pressure conditions and the hot combustion products (gases) are directed through the turbine where they

FIGURE 5.3 Diesel generator [6].

Utilities and Support System 131

expand rapidly and result in shaft rotation. The turbine also has stages, each with a row of stationary blades (or nozzles) to direct the expanding gases followed by a row of moving blades. Shaft rotation drives the compressor to draw in and compresses more air to sustain continuous combustion. The remaining shaft power is used to drive a generator which produces electricity or provide power for other equipments like MOL etc. To optimize the transfer of kinetic energy from the combustion gases to shaft rotation, gas turbines can have multiple compressor and turbine stages.

Vibration monitoring and precise control of different gas turbine systems are essential for safe and trouble-free operation. All gas turbine generators are provided with state-of-the-art control systems (mostly the latest version of digital control systems) along with integrated vibration monitoring systems. Equal importance should be given to the maintenance of all auxiliary systems and accessories, including lube oil system, seal oil system, intake air system, and gas conditioning skid for safe and efficient turbine operation. Figure 5.4 describes the gas turbine generator used in offshore platforms.

In case of ESD, the auxiliary or emergency generator takes over. These generators are driven by diesel engines, and their capacity is normally between 1.0 and 1.2 MW. More than one electrical control centers are provided on offshore installations for operating different equipment and systems. An uninterrupted

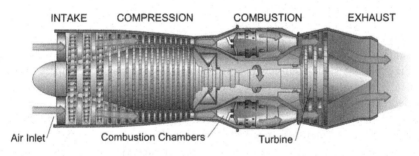

FIGURE 5.4 Gas turbine generator [7].

power supply (UPS) system supplies power to the instruments/controls along with some emergency lighting during generator shutdown and total power failure. Normally, it is designed for approximately 8 hours of supply. The UPS also has a battery bank.

5.3 INSTRUMENT AND UTILITY AIR SYSTEM [5,8]

There are normally two compressed air systems on offshore platforms – the utility and instrument air systems.

5.3.1 AIR COMPRESSORS

Normally, multistage reciprocating air compressor is used on offshore platforms to cater to the large demand of compressed air for different applications in instrumentation and utilities. These air compressors compress air to a predetermined level, normally to 10 bar (1 bar = 100 kPa), to meet the pressure requirements of different systems. Compressed air is sent to the main air receiver, the utility air receiver. Moisture present in compressed air makes it unsuitable for use in different pneumatic and electronic instruments used for control on the platforms. An air drier is used to remove moisture from compressed air. Maintaining a predetermined air pressure in the instrument air header is essential; a drop in air pressure in the header may result in platform shutdown. All pneumatic tools and winches are operated by utility air.

5.3.2 INSTRUMENT AIR AND UTILITY AIR SYSTEMS

Instrument air is stored in an air receiver. The flow and pressure in the header is maintained through a PCV. All pneumatically operated instruments, loops, and panels are supplied through this header. In case of pressure loss, a pressure switch actuates and the standby compressor is automatically loaded. Instrument air is used for PCVs, LCVs, fire loops, ESD loops, etc., and is the most important system in the installation. Instrument air pressure may be reduced to the proper pressure setting as per the system requirements before being sent to instrument air distribution depending on the pressure rating of the instrument air users.

Utility air is used for potable water system, utility generator, pedestal crane, hose reels at different locations on the decks, and many others. The start-up air for the firewater pumps is also supplied by the utility air header. A dedicated compressor is also provided for this purpose. These are "V"-type reciprocating compressor, one running and one standby. In low-pressure conditions, the standby compressor automatically begins operation. The system comprises air compressor, instrument air receiver, utility air receiver, pre-filters, dryers, and after-filters. Atmospheric air is filtered before and after it is compressed. The compressed air is stored in the utility air receiver. Normally, there are two dryers with one always in operation, which automatically changes after a certain number of hours. A unit not in operation is dried using electric heaters.

Utilities and Support System

5.4 HOT OIL SYSTEM [5]

Hot oil is an organic liquid with a very high boiling temperature and is used as an indirect heating media. In general, the following units in the processing platform need to be heated. Normally, oil is heated to the required temperature by utilizing waste heat recovered from gas turbines. Sometimes, an electrical heater or combination is also used, according to the requirement.

5.4.1 CRUDE OIL HEATER

The crude oil from the separators should be heated to aid in the demulsification process. The thermal energy of the hot oil is consumed in heating the Crude. The temperature is approximately 60°C.

5.4.2 CHEMICAL TANK

Pour point depressants sometimes gets congealed if a temperature of approximately 50°C is not maintained. Heat consumption is comparatively less in this case.

5.4.3 GLYCOL REBOILER

Rich glycol needs to be stripped of water before being recirculated. Rich Glycol is passed through stripping column in presence of stripping gas after getting heated in reboiler. This results in removal of water from Rich Glycol. The regenerated lean glycol is stored in tank below reboiler. This is a major consumer of hot oil. The high-temperature header ensures heating of the glycol reboilers. The reboiler temperature is maintained at approximately 204°C.

5.4.4 SKIMMER VESSEL

The crude oil from the closed drain header and sump caisson is sent to the skimmer vessel. There is a high possibility that this crude oil can congeal or become semisolid. In such a scenario, removing liquid from the skimmer will be difficult; therefore, the temperature is maintained around 50°C using hot oil.

5.5 POTABLE WATER SYSTEM [5,8]

The potable water system comprises potable water storage and transfer, potable water purification and distribution to consumers, topsides and hull utilities, topsides eyewash and safety showers, and hot and cold water for personnel usage. The potable water storage and transfer system is designed for filling of freshwater tanks, from bunker stations, and, from freshwater generators. It includes water makers, filters, pressurized vessels, and storage tanks. Normally, reverse osmosis (RO)-type water makers are used for generating potable water, consisting of the following:

- Sand filters
- Cartridge filters

134 Offshore Operations and Engineering

- Chemical feed
- High-pressure pump
- Permeator

The RO unit separates suspended and dissolved solids from raw sea water to make it potable. RO is a membrane process that removes 95%–99% of all dissolved minerals, 95%–97% of dissolved organic material, and 98% of biological and colloidal matter. The flow is reversed by applying pressure to sea water (feed side), which is more than the osmotic pressure. Consequently, only pure water flows through the membrane. The sand filter or diatomaceous earth (DE) filter is used to remove suspended solids. The cartridge filter is an additional protective measure to ensure filtration of particles passing through the DE filter. The high-pressure pumps are used to increase pressure of the sea water in the permeater. The RO module or permeater is a high-pressure fiber glass vessel housing the polymeric material that acts as the membrane. Prior to filtration, the feedwater is dosed with chemicals like H_2SO_4 or pH adjusters and coagulants for agglomerating the suspended matter. Thereafter, the solids are filtered out in the DE filter and are further polished using the cartridge filter.

The potable water thus produced is chlorinated before storing in the potable water storage tanks. As the membranes are highly susceptible to chlorine and the raw sea water contains chlorine, $NaHSO_3$ is also injected to remove free chlorine from the sea water. In addition, anti-scalant is also dosed to prevent the deposition of $CaCO_3$ and Mg_2CO_3 on the membrane. The produced water should contain less than 500 ppm of total dissolved solids.

5.6 WATER COOLING SYSTEM [5,8]

The seawater service system provides sea water to the central freshwater coolers, hot oil pumps, glycol booster pumps, glycol recirculation pumps, freshwater makers, inert gas generators, deck water, firewater pumps, hypochlorite generators, sewage treatment systems, and hydraulic power unit (HPU) coolers. Sea water is used to cool the condenser for the refrigeration unit, which in turn cools the closed freshwater cooling circuit.

The cooling water tank is an atmospheric vessel fitted with float-type-level control valve to maintain the water level in the tank. The cooler is of finned type and is cooled by a fan driven by electric motor.

5.6.1 FRESHWATER COOLING SYSTEM

The central freshwater cooling system is a closed loop circulating system. The system uses a refrigeration system with sea water and the evaporator cools the fresh water, thus providing fresh water at a temperature lower than cooling temperature of the condenser at the seawater temperature. Central cooling freshwater system is predominantly used by heating ventilation and air conditioning (HVAC) systems. Diesel-driven emergency generators and instrument air compressors normally have their own fan cooling water to air exchanger since they must run in emergencies.

Utilities and Support System 135

Diesel-driven main fire pumps usually use a slipstream of firewater for all cooling operations. Large kilowatt HPUs may have freshwater cooling systems that cool against the seawater system. Fresh water is used to limit contamination of the hydraulic fluid as well as damage to hydraulic parts in case of a leak from the cooling system to the hydraulic fluid.

5.7 UTILITY WATER SYSTEM [5,9]

Utility water system consists of a seawater lift pump to lift sea water and filtration unit to filter sea water used for the following:

- Water injection
- Heating, ventilation, and air-conditioning (HVAC)
- Living quarters ablutions
- Drilling facilities
- Freshwater generator
- Fire water ring main pressurization facility
- Biofouling control unit
- Sewage treatment system
- Sand jetting system
- Coarse filter backwash
- Cooling for the cooling medium system

Washdown facilities

After lifting and filtration, anti-fouling treatment, as listed above, is given to sea water before use.

5.8 DRAIN HEADER AND SUMP CAISSON [5]

There are basically two types of drains:

- Deck drain/Open drain
- Closed drain for hazardous waste

Deck drain/Open drain

Deck drains are used for dumped dirty water and other spillages from the desk into the sump caisson. Deck drains are open drains connected through the header.

Closed Drain for hazardous waste (These drain may be designed as per pressure requirement as explained below).

The closed drain header is a low-pressure header, and liquid from these drains is directed to low-pressure/high-pressure closed drain drums. Liberated gas from drums is sent to flare and liquid is sent back to low-pressure separator for retreatment.

Liquid from high-pressure vessels such as gas compressor, dehydration system, and KOD is drained to surge tank through a high-pressure drain known as a condensate drain.

The sump is an atmospheric tank designed as a gravity differential oil–water separator to treat oily water. It is an open casing about 50 m long and welded to the tubulars of the structure which have baffles. The oil droplets settle on the top and form a layer, which can be removed by vertical electrical submersible centrifugal pumps or can be lifted using gas.

5.9 HEATING, VENTILATION, AND AIR CONDITIONING EQUIPMENT [5]

Different control rooms on the platform have advanced and latest instrumentation and control systems. These components/equipment require conditioned air at controlled temperature with humidity control for efficient running of the system. At the same time, it should be comfortable for people working and living there. HVAC is provided on all offshore installations to cater to the abovementioned requirement. Normally, water-cooled heat exchangers are used.

5.10 COMMUNICATION SYSTEM [5,10]

There are a number of offshore communications technologies, and some of them are discussed below.

5.10.1 Satellite

A very small aperture terminal (VSAT) is used in offshore facilities for uninterrupted communication between offshore and onshore via satellite. An onshore terminal is known as a hub. This can be used on both fixed and floating installations. VSAT has limitations with respect to bandwidth. Figure 5.5 displays a satellite installation on an offshore facility.

FIGURE 5.5 Satellite installation on an offshore facility [11].

5.10.2 Microwave Telecommunication

Wavelengths between 1 m and 1 mm are used in microwave telecommunication for data transmission between offshore and onshore. Distance is a constraint in microwave data transmission.

5.10.3 Optical Fibers

Optical fibers are used for transmitting control signals as data between different subsea components, including subsea tree, manifolds, and jumper, as well as to control offshore and/or onshore installations. Optical fibers can support many propagating paths, thereby carrying multiple data in a single fiber. This technology has been a big support in deep sea development (Figure 5.6).

5.10.4 Cellular Services

Cellular services can be accessible at some offshore locations. Specifically, the Gulf of Mexico has cellular towers installed offshore that allow cellular communications from rigs and platforms near the coast to onshore locations, allowing offshore workers to communicate through their own private cell phones.

5.11 DIESEL SYSTEM [6]

Diesel system is designed to receive diesel on platform/Installation from Supply boat for providing fuel for the back-up power generation system.

Diesel bunkering or transfer from supply boat is a continuously manned operation, and bunkering is done with flexible hose equipped with breakaway coupling in the event of a line break due to any reason or exigencies on installation. Diesel system consists of the following:

1. Filtration system to remove impurities such as water, associated salts, and particulates to meet different user requirements.
2. Pumping system to receive and send diesel to and from different user stations.

It also provides fuel for the following systems:

- Cranes
- Life boats
- Fire water pump

FIGURE 5.6 Umbilical containing fiber communications cables [12].

5.12 SEWAGE TREATMENT SYSTEM [5,8]

There are two types of sewage system in offshore platforms – black water system and grey water system.

The grey water system is drainage water from showers, wash basins, and sinks in living quarters as well as from galleys and laundries. Grey water from hospital areas is connected to the black water system with the sewage treatment units.

The black water system collects sewage from toilets in the living quarters by a vacuum system and sends it to sewage treatment units. Showers, wash basins, and toilets in the hospital area are connected to separate headers routed to the main header at the vacuum unit. For units in most locations, the sewage treatment package must be approved or certified by the government controlling the water where the unit is located.

5.13 MATERIAL HANDLING [5]

The following equipment are normally available on platforms for moving material and equipment as well as assisting in maintenance:

- Pedestal crane
- Electric monorail hoists
- Manual hoists
- Manual trolley hoists
- The cranes transfer personnel equipment and supplies between barges or supply boats and platforms

Figure 5.7 Crane facility utilized for offshore operations.

FIGURE 5.7 Crane [13].

Utilities and Support System

5.14 OFFSHORE LOGISTICS

5.14.1 AIR LOGISTICS

Transportation of crew members and amenities is very difficult in offshore installations. A helicopter is used to transfer manpower to offshore installations. Helipad of adequate size and approved type is normally provided on offshore facilities/installations to receive helicopters in favorable weather conditions. Normally twin engine helicopter is used for offshore operations. Such helicopters are normally fitted with flotation mechanisms to avoid immediate submergence in water in case of a crash at sea [1] (Figure 5.8).

5.14.2 SEA LOGISTICS [15]

Offshore vessels are ships that specifically serve operational purposes such as oil exploration, production, construction, and support services at high seas. These vessels or ships provide logistical support to offshore operations and various facilities from construction to supply and maintenance of subsea assets. Some of the services are presented as illustrations.

Below is a brief discussion of the type of vessels and the services provided by them.

1. Offshore Support Vessel or Platform supply vessel (OSV/PSV: To provide logistical support for material, personnel, and other resources and supplies to offshore facilities. To provide services related to search and rescue during exigency.

FIGURE 5.8 Helicopter used in offshore rigs [14].

2. ESVs (emergency support vessels)/Emergency response rescue vessel: To provide immediate support during any exigency, for example, man overboard or emergency evacuation.
3. Anchor handling tug vessel (AHTV): To handle anchors and moors.
4. Tug vessel: To provide support during shuttle tanker offloading, rig movement (also performed by AHTV), etc.
5. Multisupport vessel (MSV): This is basically a PSV. These vessels are equipped with crane, diving, and/or remotely operated vehicles (ROVs). Diving and ROVs are used for inspection, maintenance, and repair/replacement of items at subsea.
6. Well intervention vessel: These are specially designed vessels with mast for hoisting and lowering of tubings and other well items for well maintenance or repair.
7. Accommodation ships/Accommodations barges: To provide temporary accommodation in offshore facilities. Sometimes PSV/OSV is also used for the same purpose.
8. Pipe laying vessel/Barge: These are specially designed vessels for laying pipeline at subsea.
9. Construction vessel/Heavy lift vessel: These are specially designed vessels for construction works, including jackets, topside installation, and subsea installations.
10. Seismic vessel: To map geological structures.

Note: Most of these vessels have dynamic positioning systems for station keeping (Figures 5.9–5.12). Normally Dynamic Positioning with Level 2 ie DP-2 vessels are preferred due to their better station keeping facilities.

FIGURE 5.9 Platform supply vessel [16].

Utilities and Support System 141

FIGURE 5.10 Supply vessel [15].

142   Offshore Operations and Engineering

FIGURE 5.11   Rescue vessel [17].

FIGURE 5.12   Pipe laying vessel [18].

REFERENCES

1. *living-on-an-offshore-oil-rig*. Available at www.nesgt.com/blog/2016/08/living-on-an-offshore-oil-rig. Accessed March 2018]
2. *Living room*. Available at www.pinterest.com/pin/362539838733921452.
3. Laundry Equipment, *Dan Marine Galley Equipment*.
4. *Food in Offshore*. Available at www.tanzaniapetroleum.com
5. *ONGC, in Graduates's Guide to Offshore Operations, 2005*.
6. *Diesel generator*. Available at www.eneria.fr/en/references/moho-nord-fpu-emergency-diesel-generator-package-2/

Utilities and Support System 143

7. *Two SGT-A30 RB gas turbines (formerly Industrial RB211) by Dresser-Rand, a business of Siemens Power and Gas which is supplied to the Penglai oil field platform in China.*
8. H. Zhang, *"An overview of marine & utility system for offshore platforms".*
9. *Azeri, Chirag & Gunashli Full Field Development Phase 3 - Environmental & Socio-economic Impact Assessment.*
10. *"How do offshore communications work?"* Available at www.rigzone.com
11. *Satellite Installation on an Offshore Facility.* Available at www.slb.com
12. *JDR cable system.* Available at www.jdrcables.com/
13. *Crane.* Available at www.haitaicrane.com
14. *Helicopter used in offshore rigs.* Available at www.youtube.com/watch?v=GLPq4f NrSkI.
15. *Supply vessel.* Available at https://products.damen.com/en/ranges/platform-supply-vessel/ psv-3300-cd.
16. *Platform supply vessel.* Available at https://freerangestock.com/photos/56014/offshore-supply-vessel.html
17. *Rescue vessel.* Available at www.errva.org.uk/
18. *Pipe laying vessel.* Available at www.knudehansen.com/references/offshore-offshore-wind-vessels/technip-pipe-lay-vessel/

6 Deep Sea Development

Oil exploration and development in deep waters has been gaining importance worldwide in the past few years. Deepwater exploration has resulted in considerable success, which has further encouraged companies to venture in this high-cost, high-risk field development with new and cost-effective technologies. We saw in Chapter 1 that the industry classifies offshore oil and gas operations according to water depth because challenges increase with increasing depth. However, technological development has been overcoming such challenges significantly. By and large, depth ranging from 300 m to 1,500 m are considered deep water, whereas depth beyond 1,500 m are considered ultradeep in the current scenario.

6.1 FACTORS DRIVING DEEP SEA DEVELOPMENT

- Future oil demand will remain strong
- Deep water has become an important prospect for big reserves
- Innovative technologies will allow economical development and reduced risks levels
- Government help in form of favorable fiscal policies
- Deepwater triangle (West Africa, Latin America, North America) likely to dominate production volume over the next decade

6.2 DEEP SEA DEVELOPMENT OPTIONS

Deep sea installations/platforms are mostly floating-type structures as it is not economical to construct and install a fixed platform at greater water depth with the available resources. Although it may be possible to locate a limited number of wells on the platform, if the number is large, subsea well remains the only option. In deep sea, mostly subsea wells are drilled and tied back to suitable installation, either in deep sea or to a shallow water platform or onshore. This depends on technoeconomical analysis (Figures 6.1 and 6.2).

While the selection of floating platform based on different parameters has been discussed in Chapter 2, it is important to understand the following few key parameters, which influence the production system design in deep water (Figure 6.3).

6.2.1 RECOVERABLE RESERVES [2]

In assessing the worth of a newly discovered deepwater field, its recoverable reserve plays a major role and is often used as an initial indicator of the field's economical feasibility.

145

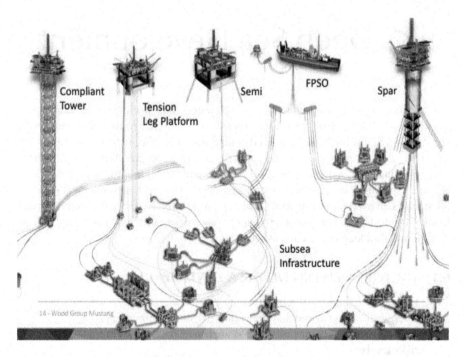

FIGURE 6.1 Deepwater production facilities. (Source: Wood Group Mustang [1].)

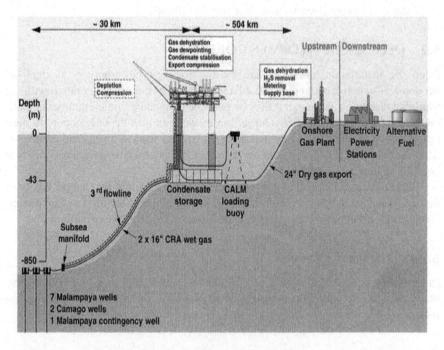

FIGURE 6.2 Subsea wells in deep water tied to fixed platform in shallow water [1].

Deep Sea Development

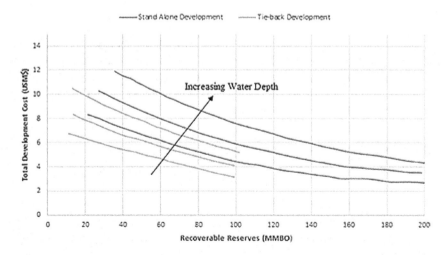

FIGURE 6.3 Recoverable reserve versus development cost [2].

6.2.2 Water Depth

Subsea oil field developments are usually divided into three categories to distinguish between the different facilities and approaches needed according to the water depth [3]:

- Shallow water subsea field development for water depth <300 m.
- Deepwater subsea field development for water depth ranging between 300 and 1,500 m.
- Ultra-deepwater subsea field development where the water depths are >1,500 m [4].

6.2.3 Challenges in Subsea due to Water Depth

Water depth governs the kind of mooring system, installation, and production risers. With the increase in water depth, there will be an increase in the difficulty of operations due to the following factors:

- Winds
- Waves
- Currents
- Pressure of water column and buoyancy
- Ice loadings

6.2.4 Production Rate

Production estimates over a field's life are also critical as rapid fall-off of rates (with low resultant ultimate reservoir recovery) can render a prospect uneconomical, even with high initial rates.

6.2.5 RESERVOIR STRUCTURE

A highly fractured reservoir will require more drainage points than a homogeneous reservoir. Low permeability coupled with low radial permeability will result in lower well flow rates requiring more wells in the field. High vertical permeability may induce water coning effect or gas breakout early in the field's life. The well system depends on the structure of the reservoir.

6.2.6 RESERVOIR PRODUCTION CHARACTERISTICS

The main production characteristics that affect system selection include:

- Flowing well temperature
- Flowing well pressure
- GOR – gas oil ratio of produced fluid from Well measured at surface
- Productivity index (PI)
- Fluid properties
- Production functions (life of field gas, oil, water production rates)

Reservoir pressure, for example, has a direct impact on well production capacity, as well as on the practical offset distance of a subsea tieback system.

Flowing well temperature and pressure along with subsea environmental condition will affect the formation of hydrates and/or wax deposition, inviting several other flow assurance problems.

Well production rates can change dramatically throughout the production life of a reservoir. This leads to difficult tradeoffs between larger flow diameters to decrease initial rate pressure drops and smaller flow diameters to improve low-rate flow performance (liquid holdup, flow stability) and fluid heat retention in late field life.

6.2.7 ENVIRONMENTAL AND GEOLOGICAL CONDITIONS

Environmental parameters to be considered include:

- **Weather conditions** impact the selection of the type of offshore installation or the decision of tying back directly to onshore. It also affects mooring system designs. For instance, weather can affect the selection of turret mooring versus spread mooring for an FPSO. Moreover, a tradeoff may have to be made whether the FPSO should be permanently moored, or a detachable or dynamically positioned system would reduce capital cost enough to offset the increased downtime associated with the periodic mooring disconnects.
- **Soil conditions** affect the mooring and foundation design of the various surface facilities. Poor soil conditions affect the cost and feasibility of the TLP's tendon foundations. It also affects the stability of the production system used. For example, the extremely hard clay soil of the North Sea bottom provides fine support for gravity-based structures. In contrast,

Deep Sea Development

the under-consolidated, soupy clay soil in the Gulf of Mexico would have platforms slipping and sliding around if they were not nailed down with deep-driven piles.

- **Loop currents** and other extreme current events lead to high loads surface-piercing structures such as a SPAR. The size and cost of mooring a SPAR in these conditions may offset other benefits of SPARs.

6.2.8 EXISTING INFRASTRUCTURE

Existing infrastructure in close proximity is always a major positive contributor in developing a complete exploration and exploitation scheme.

Key parameters influencing floating production system are summarized in Table 6.1.

FPSO in combination with shuttle tanker for oil export plays a dominant role in deepwater development. Relative distribution (by platform type), and their distribution is shown in Figure 6.4.

Distribution of these facilities in deepwater worldwide is shown in Figure 6.5.

6.3 SUBSEA FIELD DEVELOPMENT

The subsea field development prospects are characterized by large investments, tight time schedules, and the use of technology in unproven/difficult conditions. The main objective of subsea field development is to effectively maximize economical profit from offshore environment using the most reliable, safe, and cost-effective solutions available at the time of development. It is becoming commonplace in the sector of deep and ultradeep waters. Selecting the right development systems

TABLE 6.1
Floating Production System

Key Parameter	Process Platform	Processing Equipment	Mooring System	Fliser System	Well Locations	Manifolding Arrangement	Subsea Rowlines	Export System
Water Depth	×		×	×		×	×	×
Reserves	×							
Field Production	×	×		×		×	×	×
Field Life	×		×	×		×	×	×
Number of wells	×			×	×	×	×	
Fluid Properties		×		×	×	×	×	×
Reservoir Area					×	×	×	
Reservoir Depth					×			
Reservoir Pressure		×		×		×	×	
Environment	×	×	×	×				×
Infrastructure	×							×

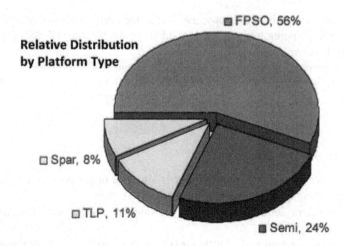

FIGURE 6.4 Relative distribution of production facilities by platform type. (Source: Wood Group Mustang [1].)

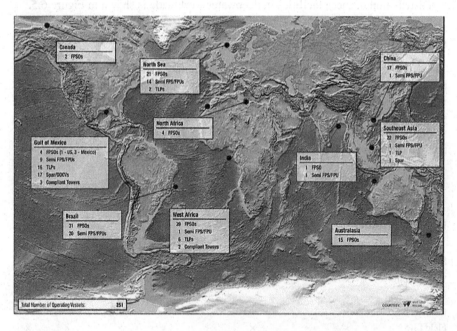

FIGURE 6.5 Deepwater production facilities worldwide – 2014. (Source: Wood Group Mustang [1].)

involves assessing the physical circumstances – water depth, reservoir configuration and location, access to oil and gas transportation – as well as the constraints placed by the local government and the institutional preferences of the investing operator. Major issues required to be considered for subsea field development are discussed as below.

Deep Sea Development 151

FIGURE 6.6 View of offshore structures present in subsea [1].

6.3.1 Subsea Well Completion

Subsea wells are located at the bottom of seabed and are not easily accessible. These wells are designed in a way to minimize reproach for intervention, either with the help of a diver or an remotely operated vehicle (ROV)-assisted system and methods in deep and ultradeep situations (Figure 6.6).

We need subsea wells for the following reasons:

- Produce deepwater or marginal fields
- Extend the life of a platform
- To produce marginal fields to existing platforms
- Producing field extremities that can't be reached by directional drilling from an existing platform

6.3.2 Subsea Christmas Tree

Subsea Christmas tree is of both dry and wet types based on its contact with seawater and is designed to be operated and monitored remotely.

6.3.2.1 Dry Tree Systems

Dry Christmas tree is installed at the production deck, and consists of individual wells from the target reservoir connected directly from the wellhead on the seafloor to the topside facility via individual flowline risers, with the Christmas tree physically located on the topside facility/platform. Dry tree systems are typically used in water depths ranging from 500 ft to 5,600 ft (shallow to deepwater) and are suitable for drilling locations such as platforms (TLP, SPAR) that can reach the locations of all the wells. Deepest dry tree facility: Devil's tower, SPAR in 1,710 m (5,610 ft) WD, GoM 2004 (Figures 6.7 and 6.8).

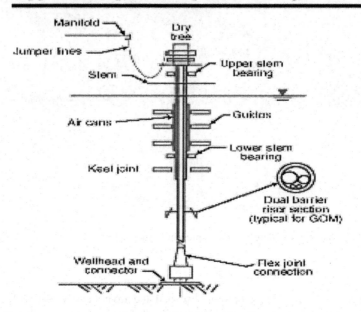

FIGURE 6.7 Schematic of dry tree system.

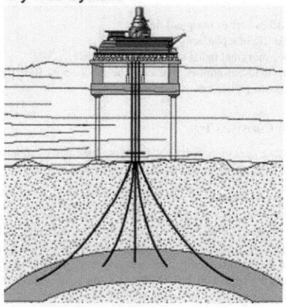

FIGURE 6.8 Side view of dry tree system [4].

Deep Sea Development 153

6.3.2.2 Wet Tree Systems

Wet tree systems are the most commonly used subsea Christmas tree currently because the tree is completely submerged in seawater and directly contacts seawater. Wet trees are typically used in conjunction with FPSOs, which represent approximately 70%–80% of global floating facilities. Wet trees can be located anywhere in a field in terms of cluster, template, or tieback methods. In deepwater fields, the wet tree system normally utilizes a remotely controlled subsea tree installation tool for well completions, which is installed with the help of Remotely Operated vehicle (ROV), however, for shallow water the diver can assist with installation and operation [4].

- Wet tree systems are segregated from production operations; the operator can predrill and complete wells faster.
- Suitable for widespread reservoir structures.
- Relatively simple, and because of the anti-corrosion performance of metal materials and the technology of remote control from the surface to subsea, they have been continuously developed in recent years.
- More than 70% of the wells in deepwater developments worldwide use wet tree systems.
- Requires fewer risers compared to dry tree systems because subsea flowlines are grouped together and routed into a limited number of risers.

For a wet tree system, the subsea field layout is usually of the following two types:

- **Subsea wells clusters:** A subsea cluster of wells gathers the production from the nearby subsea wells or from remote subsea tieback which connects to an existing platform or infrastructure through FPSO or FPU in the most efficient and cost-effective manner.
- **Direct access wells:** In marginal field development, direct access wells are applied. Direct access provides cost-effective access to the wells from the surface for workover or drilling. These developments are based on semi-submersible floating production and drilling units (FPDUs) with oil export either via pipeline or to a nearby floating storage and offloading (FSO) unit.
- For both subsea well clusters and subsea direct-access wells, the three main riser options are vertical top-tensioned risers, steel catenary risers, and flexible risers [4] (Figure 6.9).

6.3.3 Subsea Tieback Development

Tieback refer to the connection of producing wells to floating vessel or platform through risers system or directly to onshore through pipeline system. This is done after drilling small or marginal fields, after which all the subsea equipment on the seabed are installed [5].

6.3.3.1 Challenges

The main technical challenge in subsea tieback development is the design of a robust production system to economically transport the production fluids from the subsea

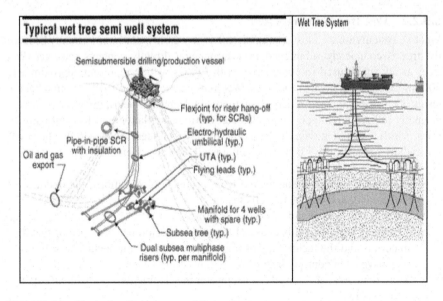

FIGURE 6.9 Typical well tree system [4].

wells to topside facilities and further to on land for processing. It requires mitigating or minimizing all flow assurance and operational risks. As subsea developments are mostly in deepwater or ultra-deepwater regions, the more challenging factors faced by operators are the locations of tieback facility, cost, and its feasibility.

The limitations in long subsea tieback system involves the limitation of flow assurance due to hydrate formation-induced plugging of the flowline as a result of a decrease in temperature along the flowline and loss of heat to the environment. Because of flow assurance problems the tieback distance becomes a limiting factor due to multiphase hydrocarbon flow in the pipeline (Figure 6.10).

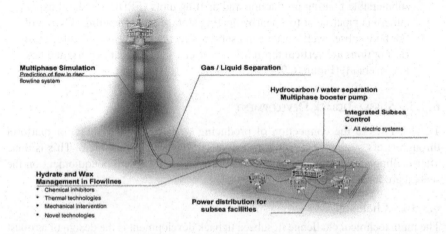

FIGURE 6.10 Subsea tieback system (tieback to Spar).

Deep Sea Development

FIGURE 6.11 (a) Tieback to floating facility, (b) Tieback to TLP, (c) Tieback to onshore facility.

The subsea development with tiebacks can be categorized as follows:

a. Tieback to floating production unit
b. Tieback to fixed platform
c. Tieback to onshore facility (Figure 6.11)

6.3.3.2 Stand-Alone Development

In stand-alone field development, a new host platform or infrastructure is constructed for the development of the subsea fields. The utilization of existing infrastructures is also considered in such a field development to minimize cost. For a stand-alone field development, the following issues are taken into consideration.

6.3.3.3 Well Groupings

When the reservoir is mapped and the number of wells is determined from the reservoir model created, the types of wells and their grouping scenario and locations are determined. Subsea wells may be installed individually, in clusters, or in a template. Wells are connected to host facility either on onshore or offshore through a manifold for transporting produced fluid. Typically, wells are grouped as satellite well system, template, clustered well system, and daisy chain.

Production field and site optimization including number and type of wells is done after reservoir mapping and modeling. Field specifications and requirement along with operator's approach play a significant role in the selection and configuration of subsea and associated surface equipment.

6.3.3.4 Satellite Well System

Subsea oil and gas fields are expanded throughout their lifecycle by single satellite tieback, while in other cases with significant new developments. A satellite well is an individual subsea well. The initial field development often includes, to a varying degree, preparation and design for some expansion. Some reasons for introducing satellite tiebacks are:

- Initially catered for to increase or maintain production rates at a certain point in time. Typically, to utilize available processing capacity after production is off plateau or to access anticipated hydrocarbon accumulations in the vicinity of the main development.
- Unplanned satellite tiebacks may be the result of lower well productivity than expected or potentially due to unforeseen reservoir compartmentalization.
- Area maturity. Typically, the smaller hydrocarbon pockets will be developed last in an area, thus in some cases, resulting in a timeframe challenge with respect to the existing infrastructure (Figures 6.12 and 6.13).

6.3.3.5 Template Well System

A subsea template is a large steel structure used as a base for various subsea structures such as wells and subsea trees and manifolds. Multiple wells can be drilled from one template. Both horizontal and deviated wells can be drilled from such a template. Normally, this formation is used to cover a large area of reservoir from one location by combining horizontal and deviated wells. This system results in saving in

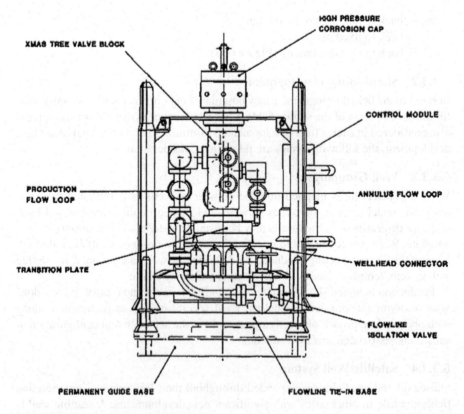

FIGURE 6.12 Satellite well system [6].

Deep Sea Development 157

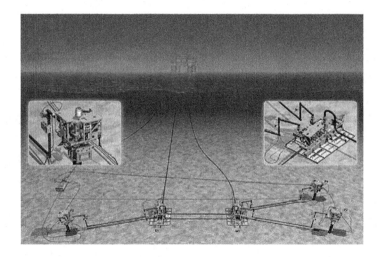

FIGURE 6.13 Satellite tieback system.

terms of umbilical and pipeline spread. Total economics depends on the combination of well cost and other infrastructure cost like umbilical, pipeline, etc. Well spacing depends on template design.

6.3.3.6 Clustered Well System

Clustered wells are generally drilled through single well templates. Individual wells are placed in a cluster formation in a close area, and, in turn, all these wells are connected to a common control station and manifold for product evacuation. Well placing is done to avoid damage caused by dropped objects to the extent possible. Umbilical and pipeline requirement in this system is governed by field layout, however, it is less than widely spaced single satellite well system (Figure 6.14).

6.3.3.7 Production Well Templates

Production well template works as the structural framework supporting manifold for fluids produced from the reservoir that flow to the wellhead and the manifold. Producing wells located in single seabed location are grouped by production well templates.

The advantages of the production well template over clustered satellite wells include:

- Prefabricated and tested jumpers and piping deployment in offshore reduces Installation time and expense.
- Provides precise locating of wells, manifold piping, and valves.

6.3.3.8 Daisy Chain

In a daisy chain configuration, wells located in different seabed locations are connected in series one after the other by flowlines. Subsea jumpers may be used to connect the flowlines to the wells, or, if applicable, the flowlines may be directly connected to the flow base of the wells. In case there are several satellite wells such as

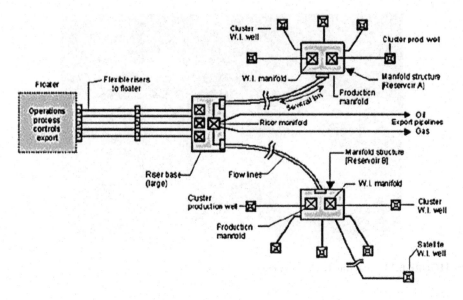

FIGURE 6.14 Clustered well system [3].

that from different marginal fields, the daisy chain field layout is considered a more economic and reliable solution compared to the cluster manifold layout. Typical features of daisy chain field architecture include:

- It may be necessary to install subsea chokes manifold on each well/Christmas tree.
- Flow assurance analysis is key to formulate the production envelop for the daisy chain flowlines, making flow assurance analysis important.
- To ensure accurate flow allocation among different wells, subsea multiphase flow meters may be required.
- Round-trip pigging is possible to minimize flow assurance by timely removal of wax build-up in the flowline [3] (Figure 6.15).

6.3.3.9 Subsea Monitoring, Control, and Communication System

Subsea monitoring, control, and communication system is the nerve center of oil and gas production from subsea developments. System architecture is complex due to the involvement of several variables and measurable parameters. A number of systems/subsystems and equipment/components is used in subsea control and communication system. Umbilical consisting of hydraulic hose, optical fiber, and electrical cable carries all the signals for control from different locations. The main control station is normally located on the topside facility and/or onshore terminal as per convenience. Subsea control and communication system is designed to meet operation, safety, environment, reliability, and flow assurance requirements.

Installation of the complete system requires careful planning and execution. Long-term maintainability and reliability may be a problem in some cases.

Deep Sea Development

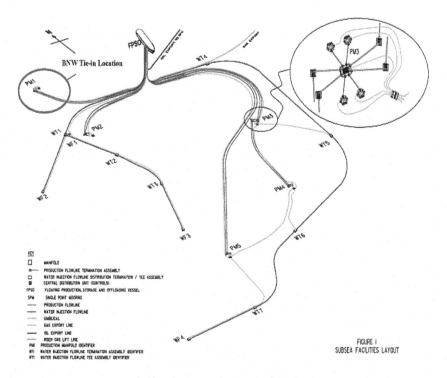

FIGURE 6.15 Manifold daisy chain layout for Bonga field development in Gulf of Guinea.

Valves, chokes, and accessories of subsea trees, manifolds, and pipelines are operated by this system. In addition, this system monitors production status by monitoring pressure, flow rate, temperature, and sand. Proper positioning of the control system results in better economics by reducing piping, cabling, and connections requirements (Figure 6.16).

Typical elements used in the system are divided into the following:

- Topside: Master/Main control station (MCS) consisting of electrical and hydraulic power unit with topside umbilical termination assembly, etc.; and
- Subsea: Umbilical along with complete subsea distribution system connecting to subsea tree/manifold as ultimate control point.

6.3.4 Main Topside Elements

Main Top side elements used in a subsea field development is described in below sections.

6.3.4.1 MCS

MCS is the heart of complete operations starting from topside to subsea for all monitoring, control, and communication between different elements. Human–Machine Interface (HMI) comprising desktop/laptop computer systems with a

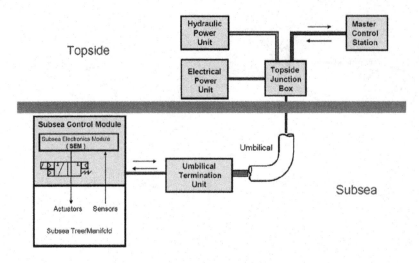

FIGURE 6.16 Subsea control system topside and subsea equipment.

suitable operating system is used for interfacing of MCS with the operator. All the subsea components and subsystems are controlled through programmable communication/data signaling system. Supervisory control network is used for all the control, monitoring, and data transmission and acquisition-related jobs. Its function includes meeting the exigencies in line with predefined parameters.

6.3.4.2 Electrical Power Unit (EPU)

Any electrohydraulic multiplexed subsea control system requires a topside unit to control and provide the necessary power to the subsea equipment. The EPU supplies dual, isolated, single-phase power for the subsea system through the composite service umbilical, together with power supply modules for the MCS and HPU (Figure 6.17).

The EPU supplies electrical power at the desired voltage and frequency to subsea users. Power transmission is performed via the electrical umbilical and the subsea electrical distribution system [6].

6.3.4.3 HPU

The HPU provides a stable and clean supply of hydraulic fluid to the remotely operated subsea valves. The fluid is supplied via the umbilical to the subsea hydraulic distribution system, and to the subsea control module (SCM) to operate subsea valve actuators [7] (Figure 6.18).

6.3.5 Topside Umbilical Termination Assembly (TUTA)

It provides the interface between the topside control equipment and the main umbilical system. This fully incorporates electrical junction boxes for the electrical power and communication cables, as well as tube work, gauges and block and bleed valves for the appropriate hydraulic and chemical supplies [4] (Figure 6.19).

Deep Sea Development 161

FIGURE 6.17 Electrical power unit (EPU).

FIGURE 6.18 Hydraulic power unit (HPU). (Source: Oceaneering [8].)

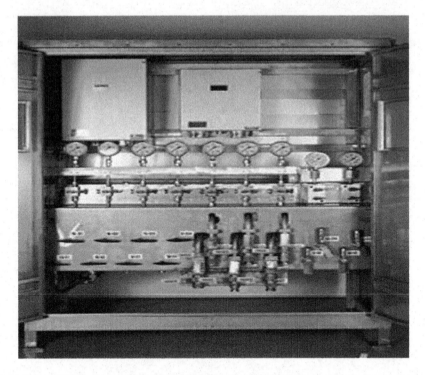

FIGURE 6.19 Topside umbilical termination assembly (TUTA) [7].

TUTA is typically located at the umbilical J-tube on the host. It includes an electrical enclosure in a lockable stainless-steel cabinet. The valves in the TUTA comply with requirements for valves in flammable services. It is designed for hang-off and includes a bull-nose suitable for pulling the umbilical up.

6.4 SUBSEA ELEMENTS [9]

- Functions of subsea distribution system (SDS):
  1. Hydraulic power distribution
  2. Chemical injection distribution
  3. Electrical power distribution
  4. Communication distribution
- Components of SDS: [9]
  - Subsea accumulator module (SAM)
  - Subsea umbilical termination assembly (SUTA): It mainly consists of inboard multiple quick connect plates, mounting steel structures, a lifting device, mud mat, logic cap, long-term cover, field assembled cable termination, and electrical connectors.
    1. Umbilical termination head (UTH)
    2. Hydraulic distribution module (HDM)

Deep Sea Development

3. Electrical distribution module (EDM)
4. Flying leads

- Subsea distribution assembly (SDA): It mainly consists of HDM and EDM. The HDM consists of inboard multiple quick connect (MQC) plates, mounting steel structures, lifting padeyes, a mudmat, logic cap, and long-term cover. The EDM consists of bulkhead electrical connectors and cables and, in some cases, an electrical transformer module.
- Hydraulic flying leads (HFL): It mainly consists of two outboard MQC plates with holding structures and steel tubes. Electrical flying leads (EFLs) mainly consist of two electrical connectors and a number of cables [9].
- EFL
- MQC
- Hydraulic coupler
- Electrical connector
- Logic caps

1. **Mudmat:** Defined as the structure whose function is as a load-bearing beam to prevent offshore construction from sinking into soft unconsolidated soil on the seabed.
2. **Padeyes:** This is a device usually with round opening projection with a plate. This is used for holding a load on deck of boat in steady condition or for lifting of items by engaging hooks. It is a kind of fairlead and often is bolted or welded to the deck or a boat's hull. It is also used in oil and gas projects to assist in lifting (Figure 6.20).

SCM: Subsea control manifold
FUTURE: Spare receptors to be used in the future

6.4.1 SDS COMPONENTS

6.4.1.1 Umbilical

It is composed of tubing, piping, and electrical conductors, as shown in Figure 6.21.

It runs from the host facility to subsea production equipment, and is used to transmit control fluids and electricity to subsea safety and production equipment. Subsea control and monitoring work in this manner.

Dedicated tubes are used to monitor pressure and inject fluids from the host facility to critical areas (subsea production and safety system) [11]. Electronic conductors are used to transmit power to operate subsea electronic devices.

6.4.1.2 Subsea Umbilical Termination Assembly (SUTA) [6]

It works as the subsea interface for the umbilical and may serve as the distribution center for the hydraulic and chemical services at the seabed. It is connected to subsea trees via HFL.

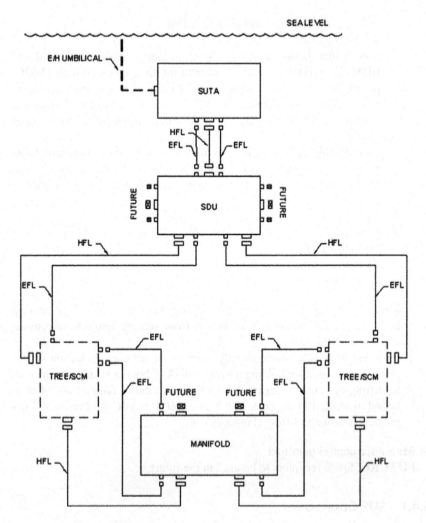

FIGURE 6.20 Subsea distribution system: block diagram (Source: www.oilandgastechnologies.wordpress.com.)

SUTA is typically composed of umbilical termination head (UTH), flying leads connect UTH and HDM, mudmat, and MQC plates; and is installed with or without drilling rig assistance [4].

The umbilical is permanently terminated in the UTH. EFLs are used to connect electrical power/communication from the UTA to the manifold and tree-mounted SCM. The UTH, HDM, and EDMs are each independently retrievable from the UTA's mudmat. Each of the electrical quads (umbilical cables with four conductors) is terminated in electrical connectors at the UTH.

The EFL interconnects between these UTH connectors and the EDM connectors, routing power and communication from the UTH to the EDM. Subsea electrical distribution is done from the EDM to the subsea trees and production manifolds [4].

Deep Sea Development 165

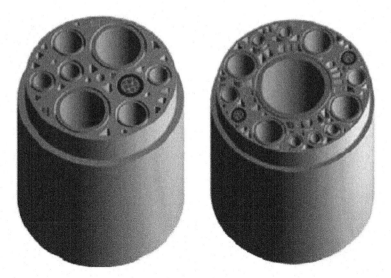

FIGURE 6.21 Subsea steel umbilical [10].

6.4.1.3 Umbilical Termination Head (UTH)

It consists of a structural frame, hinging stab, MQC plates, super-duplex tubing, and bulkhead-type ROV electrical connectors. Distribution to production equipment is done by routing the umbilical services to the MQC plates and electrical connectors.

All tubing in the UTH and HDM is welded to the hydraulic couplers located within the MQC plates. The number of welded connections between the couplers and the tubing is kept at a minimum. The design allows for full opening through the tubing and the welded area [4] (Figure 6.22).

At minimum, the UTH is designed to allow termination of a minimum of nine umbilical steel tube lines and two electrical quads. The structural frame is designed to securely attach to and support the umbilical and end termination, as well as provide mounting locations for the MQC plates and bulkhead electrical connectors.

6.4.1.4 Subsea Distribution Assembly

This distributes hydraulic supplies, electrical power supplies, signals, and injection chemicals to the subsea facilities. The facilities can be a subsea template, a satellite well cluster, or a distribution to satellite wells (Figure 6.23).

6.4.1.5 Hydraulic Distribution Manifold/Module (HDM)

It distributes hydraulic fluid (LP and HP) and chemicals to each of the subsea trees, production manifolds, and future expansions. It consists of multiple inboard MQC plates on a frame mounted on an umbilical termination assembly mudmat (Figure 6.24).

FIGURE 6.22 Umbilical termination head [4].

FIGURE 6.23 Subsea distribution assembly [8].

Deep Sea Development

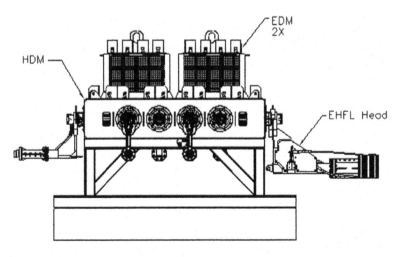

FIGURE 6.24 Hydraulic distribution unit (EDM and HDM) [12].

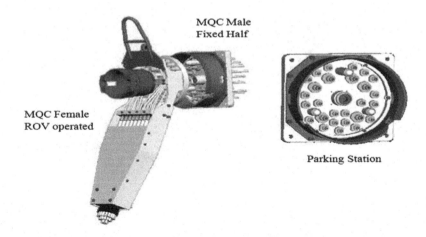

FIGURE 6.25 Multiple quick connects [13].

6.4.1.6 Electrical Distribution Manifold/Module

The number of electrical connectors in series is kept to a minimum. Redundant routing to the module connectors should follow different paths, if possible. The cables should be installed into self-pressure compensated fluid-filled hoses, and the fluid is of dielectric type.

6.4.1.7 Multiple Quick Connects

MQC plates are outfitted with an appropriate number of couplers to match the number of tubes in the HFLs. All couplers are energized, and the MQC is capable of withstanding full coupler pressures (Figure 6.25).

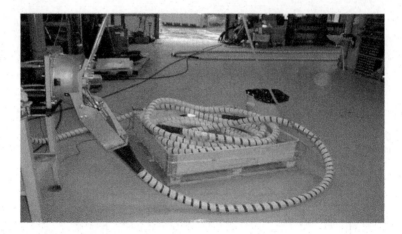

FIGURE 6.26 HFL assembly [4].

FIGURE 6.27 Hydraulic couplers [14].

6.4.1.8 Hydraulic Flying Leads (HFL)

The HFL are used for supplying hydraulic and chemical fluid for different purposes from, SUTA to tree, or, SDA to tree. HFLs are made up of three main components: the tubing bundle, the steel bracket assembly heads, and the MQCs. The tubing bundle is terminated at both ends with MQC plates (Figure 6.26).

6.4.1.9 Hydraulic Couplers

Hydraulic couplers for deepwater applications need a spring strong enough to seal against the external pressure head to prevent seawater from contaminating the hydraulic fluid, and need to be designed such that only a low-pressure force is required for makeup. A fully pressure-balanced coupling is preferable. It is important for hydraulic couplers to have adequate flow paths to ensure that adequate hydraulic response times are achievable [14] (Figure 6.27).

FIGURE 6.28 Electrical flying lead assembly.

6.4.1.10 Electrical Flying Leads

The EFL connects the EDU to the SCM on the tree. The EFL assembly is composed of a pair of electrical wires enclosed in a thermoplastic hose and fitted at both ends with soldered electrical connectors (Figure 6.28).

6.4.1.11 Logic Caps

Hydraulic and chemical distribution equipment includes dedicated MQC plates, known as logic caps, which provide the ability to redirect services by replacing an outboard MQC plate with an ROV. Logic caps provide the flexibility to modify distribution of hydraulic or chemical services due to circuit failures or changes in system requirements.

6.4.1.12 Subsea Accumulator Module

This is a subsea unit that stores hydraulic fluid such that adequate pressure is always available to the subsea system even when other valves are being operated. SAM is used to improve the hydraulic performance of the subsea control system in trees and manifolds. Basically, this will improve hydraulic valve actuation response time and system hydraulic recovery time, at minimum system supply pressures [15] (Figure 6.29).

6.4.1.13 Subsea Control Module

The SCM is an independently retrievable unit commonly used to provide well control functions during the production phase of subsea oil and gas production. Typical well control functions and monitoring provided by the SCM include actuation of fail-safe

FIGURE 6.29 Subsea accumulator module [15].

FIGURE 6.30 Typical SCM. (Source: FMC.)

return production tree actuators and downhole safety valves; flow control choke valves, shutoff valves, manifold diverter valves, shutoff valves, chemical injection valves; actuation and monitoring of surface-controlled reservoir analysis and monitoring systems, sliding sleeve, choke valves; monitoring of downhole pressure, temperature, and flow rates; etc. [4] (Figure 6.30).

Deep Sea Development 171

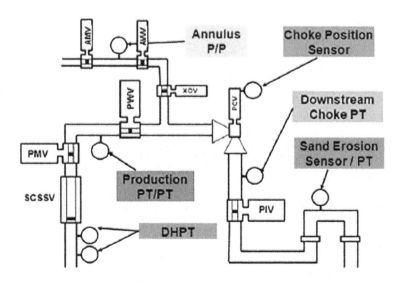

FIGURE 6.31 Subsea transducers/sensors located on a subsea tree.

6.4.1.14 Transducer/Sensor

Subsea sensors are at multiple locations on the trees, manifold, and flowlines. Figure 6.31 shows the subsea transducers/sensors (pressure and temperature) locations on a subsea tree. Tree-mounted pressure sensors and temperature sensors measure upstream and downstream of the chokes. Software and electronics in the SCMs compile sensor data and system status information with unique addresses and time-stamp validations to transmit to the topside MCS, as requested (Figure 6.31).

6.4.1.15 Subsea Production Control System

SPCS is necessarily designed in an integrated control system that works remotely (Figure 6.32).

Functions of an SPCS include opening and closing subsea Christmas tree production; annulus; subsea production manifold flowline valves, pigging valves, and crossover valves, operating SCSSVs, chemical injection valve; adjusting subsea production chokes; and monitoring temperature, pressure, flow rate, and some other data from tree-mounted, manifold-mounted, or downhole instrumentation (Figure 6.33).

6.4.1.15.1 Electrical Power System and Communication

The following main parameters need to be determined by a system power demand analysis:

- Voltage at subsea electronic module (SEM) for maximum and minimum SEM power loads;
- Voltages at each SEM at maximum and minimum numbers of SCMs on the subsea electrical distribution line;

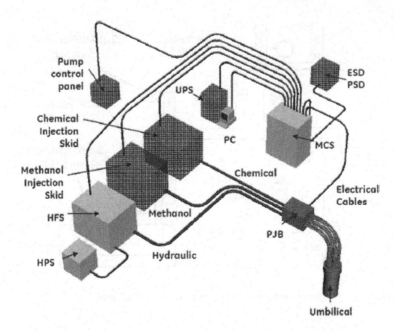

FIGURE 6.32 Subsea production control system topsides. (Source: Vetco Gray.)

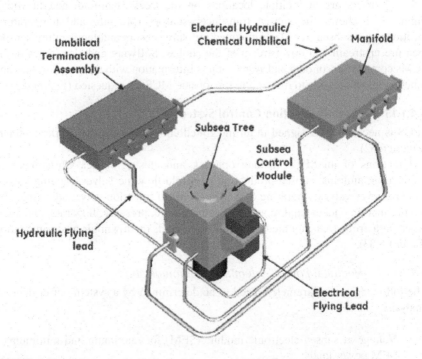

FIGURE 6.33 Subsea production control system subsea. (Source: Vetco Gray.)

Deep Sea Development

- Voltages at SEM at minimum and maximum designed umbilical lengths;
- Voltages at SEM at cable parameters for dry and wet umbilical insulations;
- Minimum and maximum subsea power requirements;
- Maximum current load.

6.4.1.16 Types of Control Systems

Monitoring, control, and communication systems have multiplex electrohydraulic/ electrical control system with data communication to make it compatible with the computer operating system. All the latest systems are operated either from local control station or linked remote stations. The basic function of such a system is flow control by controlling different types of valves in the system, as well as monitoring and transmitting data to main station for analysis and diagnostics.

The basic control system types include:

1. Direct hydraulic
2. Piloted hydraulic
3. Sequenced hydraulic
4. Multiplex electrohydraulic
5. All-electric

6.4.1.16.1 Direct Hydraulic Control System

In direct hydraulic control system, each valve is remotely controlled using its own hydraulic line. Distance becomes a limiting factor.

In this system, control valves are actuated (opened) by directly applying hydraulic pressure on the valve actuator and deactivated (closed) by removing the pressure from actuator. In other words, control valve is opened by flooding the actuator with pressurized control fluid and closed by venting fluid to reservoir/ accumulator from actuator. The principle of a direct hydraulic control system is explained bellow.

6.4.1.16.1.1 Advantages

- Minimum subsea equipment
- Low cost
- Reliability is high because the critical components are on the surface
- Maintenance access is very good because all critical components are on the surface

6.4.1.16.1.2 Disadvantages

- Response is slow
- Large number of hoses
- Limited monitoring capabilities and distance limitations due to long response time and umbilical costs
- Limited operational flexibility [4] (Figure 6.34).

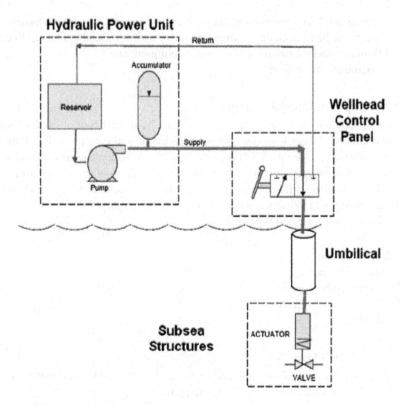

FIGURE 6.34 Direct hydraulic control system [4].

6.4.1.16.2 *Piloted Hydraulic Control System*

This system has dedicated hydraulic pilot supply for each subsea function and hydraulic supply line to each SCM. The presence of accumulator increases valve opening response time.

The difference of the system from the direct hydraulic system is that the umbilical does not include large bore hoses to achieve the performance requirements. The pilot system uses a small bore hose for the pilot line and a larger bore hose for the supply line. The hydraulic pilot volume to actuate the pilot valve is very small and there is very little volume flow required to energize the pilot valve and open the subsea valve. As a result, the valve actuation time improves [4] (Figure 6.35).

6.4.1.16.2.1 *Advantages*
- Low cost
- Reliability is high because the critical components are on the surface
- Maintenance access is good because the majority of the components are on the surface
- Proven and simple subsea equipment

Deep Sea Development 175

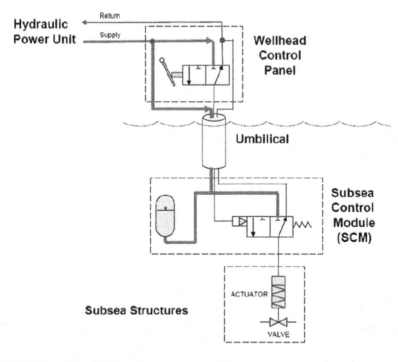

FIGURE 6.35 Piloted hydraulic control system [4].

6.4.1.16.2.2 Disadvantages
- Response is slow
- Large number of hoses
- Limitation in distance because response is slow
- No subsea monitoring because there are no electrical signals [4]

6.4.1.16.3 Sequenced Hydraulic Control System The sequenced hydraulic control system consists of several sequence valves and accumulators. Various complicated programs are designed by series-parallel for sequence valves (Figures 6.36 and 6.37).

Valves open at predetermined sequences depending upon the magnitude of the signal. The system works by adjusting the regulator up at opening pilot pressure to open valve. As we can see in Figure 6.38, the first valve opens at 1,500 psi and sequentially remains open at a difference of 500 psi pressure.

6.4.1.16.3.1 Advantages
- Improved system response compared to direct hydraulic control and piloted hydraulic control systems
- Reduced umbilical compared to direct hydraulic control and piloted hydraulic control systems
- Small number of hydraulic hoses

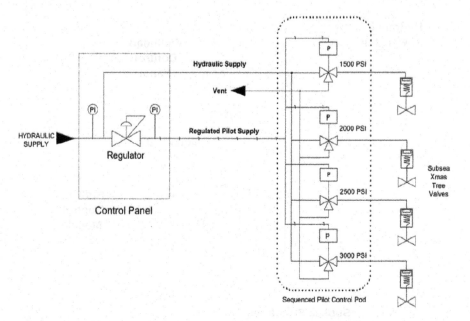

FIGURE 6.36 Sequenced hydraulic control system. (Source: TotalFinaElf.)

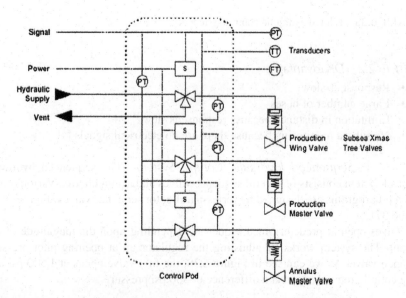

FIGURE 6.37 Multiplex electrohydraulic control system. (Source: TotalFinaElf.)

6.4.1.16.3.2 Disadvantage
- Slow operation
- The sequence of valve opening and closing is fixed
- Distances are limited because response is slow

Deep Sea Development

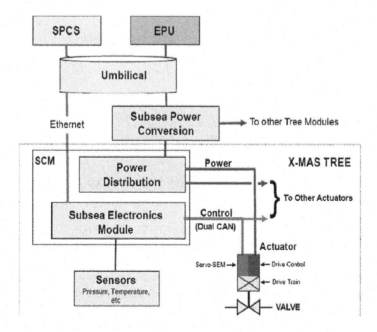

FIGURE 6.38 All electric control systems.

- No subsea monitoring because there are no electrical signals
- Increase in the number of surface components
- Increase in the number of subsea components

6.4.1.16.4 Multiplex Electrohydraulic Control System

The multiplex electrohydraulic control system is composed of master control station (MCS), a subsea electronic module. Subsea electronic module has a microprocessor which communicates with the MCS and performs MCS commanded function. It allows multiple SCM at the same communication electrical and hydraulic supply lines, implying that wells can be controlled by single umbilical from a single location.

Cost is high because of the involvement of SEM, computers, and computer software at topside. However, it can be balanced by less umbilical and advanced technology used for complex fields and long distances (more than 5 km).

When a digital signal is sent to the SEM, it excites the selected solenoid valve, thereby directing hydraulic fluid from the supply umbilical to the associated actuator. The multiplex electrohydraulic control system is capable of monitoring pressure, temperature, and valve positions by electrical signals, without further complicating the electrical connections through the umbilical.

6.4.1.16.4.1 Advantages
- Good response times over long distances
- Smaller umbilical diameter
- Allows control of many valves/wells via a single communication line

- Redundancy is easily built in
- Enhanced monitoring of operation and system diagnostics
- Ideal for unmanned platform or complex reservoirs
- Able to supply high volume of data feedback
- No operational limitations

6.4.1.16.4.2 Disadvantages
- High level of system complexity
- Increase in surface components
- Increase in subsea components
- Recharging of the hydraulic supply over a long distance
- Hydraulic fluid cleanliness
- Materials compatibility
- Limitations over long distance tiebacks [3] (Figure 6.39).

6.4.1.16.5 All Electrical Control System

All electrical control system is a fast-response control system without any hydraulic system. It consists of all electrically operated elements. Elimination of hydraulic circuits and accumulators helps in increasing the response time. Advent of long-distance subsea power transmission has made this concept easily acceptable. The principle of an all-electric control system is shown in Figure 6.40.

All valves including the choke valve are fitted with electrical actuators. SCM is supplied power from the topside facility, that is, offshore installation or from onshore facilities. System architecture becomes simpler, and response time becomes faster. It offers higher degree of flexibility in case of expansion or inclusion of new wells.

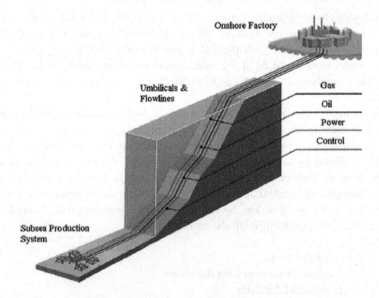

FIGURE 6.39 Power supply from onshore to subsea (Source: Vetco Gray.)

Deep Sea Development

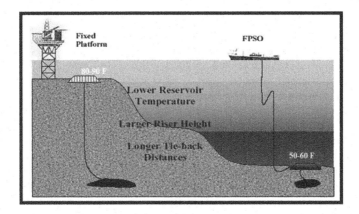

FIGURE 6.40 Shallow versus deepwater.

6.4.1.16.5.1 Advantage
1. Quick response
2. Hydraulic circuit and hydraulic accumulator eliminated
3. Umbilical without hydraulic line required
4. Easier installation due to elimination of hydraulic circuit and hydraulic accumulator
5. Adding new well becomes easier

6.4.1.16.5.2 Disadvantage
1. In case of long-distance transmission, introduction of bulky transformer for supply regulation
2. Limitation due to long transmission distance

6.5 SUBSEA POWER SUPPLY

It is a major component of subsea processing without which operations cannot continue. These include:

1. Electrical power unit (EPU)
2. Uninterrupted power supply (UPS)
3. Hydraulic power unit (HPU)

Power system supplies various types of powers to subsea equipment: SCM, SEM, pumps, motors, and actuators located on trees/manifolds and sensors. Power source may come from onshore or from the platform.

6.6 FLOW ASSURANCE

Flow assurance can be defined as the ability to produce fluids economically from the reservoir to the point-of-sale via production facility over the life of a field in any environment. Flow assurance focuses on the entire engineering and production

180 Offshore Operations and Engineering

lifecycle from the reservoir through refining to ensure, with high confidence, that the reservoir fluids can be moved from the reservoir to the refinery smoothly and without interruption. It is a rigorous engineering process that works to maximize production by ensuring unrestricted production flow path throughout the field's life with minimum lifecycle costs.

Main flow assurance driver:

- Ensures sufficient pipeline capacity
- Minimizes the risk of flowline blockages
  - hydrates and wax
  - asphaltenes, sand, and scale
- Provides the ability to remove blockages
- Minimizes downtime during operations
  - pigging
  - start-up and ramp-up
  - shut-in
  - lugging
- Enhances operability

6.6.1 Shallow versus Deep Flow Assurance Scenario

As shown in Figure 6.40, the seabed temperature in deepwater is lower than the seabed temperature in shallow environment. Moreover, the riser and tieback lengths are greater in the case of deepwater, which increases the frictional forces and the back pressure. This makes the flow assurance scenario very different in shallow and deepwater.

6.6.2 Flow Assurance Challenges

Flow assurance challenges mainly focuses on the prevention and control of solid deposits that could potentially block the flow of produced fluid.

These solids have the potential to deposit anywhere from the near wellbore and perforation to wellbore, topside surface facilities, and pipelines. Depending upon the specific fluid properties, hydrodynamic, and heat transfer characteristics of a given field development system, the pressure and temperature PT path may intersect one or all the three elements of hydrocarbon solid formation, namely, asphaltene, wax, and hydrate.; Flow path from reservoir to facility is shown Figure 6.41.

6.6.3 Troublemakers

6.6.3.1 Gas Hydrates

Gas hydrates are compounds where gas molecules like methane, ethane, propane, CO_2, H_2S are trapped within the crystal lattice structure of ice. High pressure and low temperature is required for the formation of gas hydrates. Hydrate formation during normal operations, following shut-ins, and during start-up can lead to large production losses and tricky and costly intervention problems.

Deep Sea Development

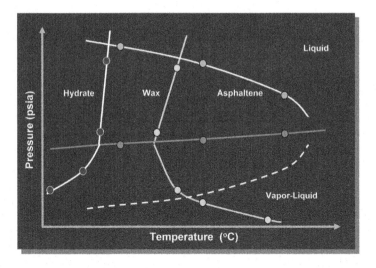

FIGURE 6.41 Stability diagram.

6.6.3.2 Paraffin/Wax
Paraffins are high-molecular-weight alkanes ($C_{20}+$) that can build up as deposits in the wellbore, feed lines, etc. These organic deposits can act as chokes within the wellbore, resulting in a gradual decrease in production over time with increase in the thickness of the deposits. They are very sensitive to temperature.

6.6.3.3 Asphaltene
Asphaltenes are high-molecular weight aromatic organic substances that are soluble in toluene but are precipitated by alkanes (n-heptane/n-pentane). Asphaltenes remain dissolved in colloidal suspension under high pressure and temperature conditions of reservoir, however, are destabilized by changes in temperature and pressure due to primary depletion and begin precipitating in the reservoir.

6.6.3.4 Scales
Scales are mineral salt deposits that may occur on wellbore tubulars and components as the saturation of produced water is affected by changing temperature and pressure conditions in the production conduit. In severe conditions, scales create a significant restriction, or even a plug, in the production tubing.

6.6.3.5 Erosion
Erosion is the wearing out of the tubing, pipelines, and flowline walls due to solid particles such as sand or due to liquid impingement at high fluid velocities.

6.6.3.6 Corrosion
Corrosion is the wearing out of flowline thickness due to the chemistry of the produced fluids it can be result of erosion also.

6.6.3.7 Slugging

Slugging is caused by instabilities of gas and liquid interface and liquid sweep-out by gas inertial effects.

6.6.3.8 Severe Slugging

At low gas and liquid flow rates, unsteady-state flow may occur in such two-phase pipeline-riser systems. The cyclic unsteady-state flow characterized by large-amplitude, relatively long-period pressure, and flow rate fluctuations is known as severe slugging.

At relatively low flow rates, liquid accumulates at the bottom of the riser, blocking the gas, until sufficient upstream pressure has been built up to surge the liquid slug out of the riser followed by gas surge. After fluid and gas surge, part of the liquid in the riser falls back to the riser base to create a new blockage and the cycle repeats. This transient cyclic phenomenon causes periods of no liquid and gas production at the riser top followed by very high liquid and gas surges, and is called severe slugging [16] (Figure 6.42).

6.6.4 Typical Flow Assurance Processes

The main part of flow assurance should be done before the front-end engineering and design (FEED) process. The requirements for each project are different and, therefore, project-specific strategies are required for flow assurance problems (Figure 6.43). Nevertheless, the approach to deal with main flow assurance issues is discussed in subsequent paragraphs for understanding of the importance of the topic in any project.

6.6.5 Fluid Characterization and Flow Property Assessments

This ensures the validity of flow assurance by carefully studying the fluid properties of the crude sample collected from the well or from the analogous sample from a nearby production well. The key analysis on the sample are:

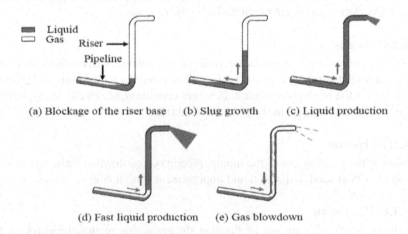

(a) Blockage of the riser base   (b) Slug growth   (c) Liquid production

(d) Fast liquid production   (e) Gas blowdown

FIGURE 6.42 Stages of severe slugging [16].

Deep Sea Development

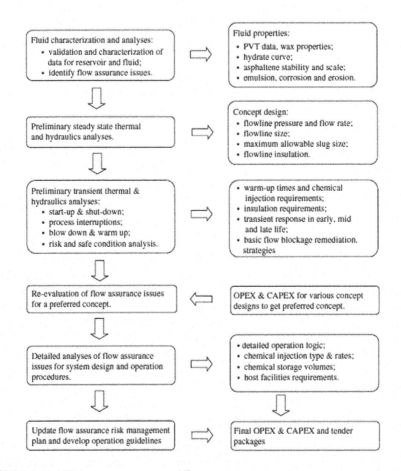

FIGURE 6.43 Flow assurance process [4].

- PVT properties: Phase composition, GOR, bubble-point pressure
- Wax properties: Cloud point, pour point, asphaltene stability
- Water salinity, either from the produced water samples or resistivity log

Based on the above analysis, hydrate stability curves are developed and methane dosing is also calculated [4].

6.6.6 Steady-State Hydraulic and Thermal Performance Analyses

This stage generates steady-state flowline models and determines the following parameters:

- The relationship between flowrate and the pressure drop along the flowline.
- Check pressure and temperature distribution along the pipeline to avoid hydrate formation.

184　　Offshore Operations and Engineering

- Riser base temperature as a function of flow rate and combined insulation system and prevent the riser temperature to go below the minimum value for cool down at the maximum range of production ranges.
- Maximum flowrate such that the temperature does not exceed any upper limits set by the separation and dehydration processes or by equipment design.

6.6.7 System Design and Operability [4]

In a system design, the entire system from the reservoir to the end user needs to be considered to determine applicable operating parameters; flow diameters and flow rates; insulation for tubing, flowlines, and manifolds; chemical injection requirements; host facilities; operating strategies and procedures to ensure smooth flow of produced fluids.

6.6.7.1 System Start-Up

During start-up, as the system is at a lower temperature, hydrate inhibitor needs to be injected. If start-up rates are higher, then only the tree needs to be treated with the inhibitor and inhibitor need not to be injected downhole. However, if the start-up rate is low, the hydrocarbon flow needs to be treated with inhibitors downhole. Once the tree is outside the hydrate region, hydrate inhibitor can be injected at the tree and the flow rate can be increased to achieve system warm-up.

The following start-up philosophies are used for cases where well start-up poses a risk for flowline blockage, particularly if a hydrate blockage is suspected based on flow assurance studies from the design phase:

- Wells will be started up at a rate that allows minimum warm-up time, while considering drawdown limitations.
- It may not be possible to fully inhibit hydrates at all water cuts, particularly in the wellbore. High water-cut wells will be brought online without being fully inhibited.
- The system is designed to inject hydrate inhibitor (typically methanol) at the tree during start-up or shutdown. The methanol injected during the initial start-up will inhibit well jumpers, manifold, and flowline if start-up is interrupted and blowdown is not yet possible (Figure 6.44).

6.6.7.2 System Shutdown

In general, planned and unplanned shutdowns from the steady state are the same, with the exception of a planned shutdown; hydrate inhibitors can be injected into the system prior to shutdown. Well shutdown also poses a significant hydrate risk. The following philosophies may be adopted during shutdown operations:

- The subsea methanol injection system is capable of treating or displacing produced fluids with hydrate inhibitor between the manifold and SCSSV following well shutdown to prevent hydrate formation.
- Hydrate prevention in the flowlines is accomplished by bringing the flowline pressure down to less than the hydrate-formation pressure at ambient seabed temperatures.

Deep Sea Development

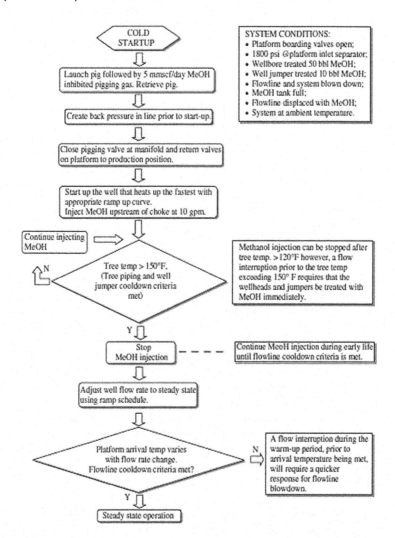

FIGURE 6.44 Operating logic chart of start-up for a cold well and a cold flowline [4].

- Most well shutdowns will be caused by short-duration host facility shutdowns. The subsea trees, jumpers, and flowlines are insulated to slow the cooling process and allow the wells to be restarted without having to initiate a full shutdown operation.

6.6.8 Transient Flow Hydraulic and Thermal Performance Analyses

Transient flowline system models have to be constructed to check any flow problems. Transient flowline analyses generally include the following scenarios:

- Start-up and shutdown
- Emergent interruptions

186   Offshore Operations and Engineering

- Blowdown and warm-up
- Ramp up/down
- Oil displacement
- Pigging/slugging

This modeling is done to ensure that the fluid temperature in the system does not exceed the hydrate dissociation temperature corresponding to the pressure at every location; otherwise, a combination of an insulated pipeline and the injection of chemical inhibitors into the fluid must be simulated in the transient processes to prevent hydrate formation, and the same needs to be done in the field.

6.6.9 HYDRATE PREVENTION METHODS

Hydrate prevention techniques for subsea systems are discussed below.

6.6.9.1 Thermodynamic Inhibitors

These inhibit hydrate formation by reducing the temperature at which hydrates are formed, that is, they shift the stability curve to further lower temperature. Hence, lower temperature are required to initiate hydrate formation. Methanol and mono-ethyl glycol (MEG) are the most commonly used inhibitors. Methanol can provide higher hydrate temperature depression.

The risks of using thermodynamic inhibitors include:

- Inadequate injection of inhibitor, particularly due to not knowing water production rates
- Inhibitor not going where intended (operator error or equipment failure)
- Environmental concerns, particularly with methanol discharge limits
- Ensuring remote location supply
- Ensuring chemical/material compatibility
- Safety considerations in handling methanol topside (Table 6.2)

Water production multiplied by the methanol dosage is the inhibitor injection rate, which changes throughout the field's operating life due to typically decreasing operating pressures and increasing water production. Hence, more the water cut, more will be the inhibitor injection rates.

TABLE 6.2
Hydrate Inhibition Comparison

Inhibitor	Advantage	Disadvantage
Methanol	Less viscous	Lower recovery due to gas and condensate phase loses.
	Less likely to cause salt precipitation	Impact on downstream processing plants
	Regeneration cost is less compared to MEG	Low flash point
MEG	Easy to recover	Higher viscosity
	Low gas and condensate solubility	Causes salt precipitation

6.6.9.2 Low Dosage Inhibitors (LDIs)

LDIs do not change the hydrate formation temperature. They either interfere with the formation of hydrate crystals or agglomeration of crystals into blockages. Anti-agglomerates can provide protection at higher subcooling temperatures than kinetic hydrate inhibitors. However, low-dosage hydrate inhibitors are not recoverable and are expensive.

6.6.9.3 Kinetic Inhibitors (KIs)

These are low-molecular weight, water-soluble polymers which prevent hydrate blockage by bonding to the hydrate surface and delaying hydrate nucleation or growth.

Unlike TDI, these inhibitors work independently of water cuts, but are limited to relatively low subcooling temperatures (<20°F), which may not be sufficient for deepwater applications. For greater subcooling, KIs must be blended with a thermodynamic inhibitor. Additionally, the inhibition effect of KIs is time-limited.

6.6.9.4 Anti-Agglomerates (AAs)

They allow hydrate crystals to form but keep the particles small and well dispersed in the hydrocarbon liquid. Moreover, they inhibit hydrate plugging rather than hydrate formation. AAs can provide relatively high subcooling up to 40°F, which is sufficient for deepwater applications. A potentially important advantage of LDIs is that they may extend field life when water production increases.

6.6.9.5 Low-Pressure Operations

Hydrate formation can be prevented by maintaining operational pressure below the pressure corresponding to the ambient temperature based on the hydrate stability curve. So, at ambient temperature of 4°C, pressure should be <300 psia, but this is highly unpractical.

By using subsea choking and keeping the production flowline at a lower pressure, the difference between hydrate dissociation and operating temperatures (i.e., subcooling) is reduced. This lower subcooling will decrease the driving force for hydrate formation and can minimize the inhibitor dosage.

6.6.9.6 Water Removal

For export pipelines, dehydration is a common technique used for hydrate prevention. For subsea, subsea separation system can be employed for reducing the percentage of water in the subsea flowlines. It will not only control hydrates but will also make the produced fluids lighter and easier to lift, eventually reducing the topside water handling, treatment, and disposal. Subsea separation is covered in detail in emerging technology.

6.6.9.7 Thermal Insulation and Heating

This serves two purposes: prevents hydrate formation control by maintaining temperatures above hydrate formation conditions, and extends cooldown time before reaching the hydrate formation temperature. There are three types of insulation listed in ascending order of cost: coatings applied to pipes; pipe-in-pipe

188 Offshore Operations and Engineering

(PIP) or bundling; and vacuum-insulated pipes or pipes with insulating gases. Insulation is generally not applied to gas production systems because the production fluid has low thermal mass and experiences JT cooling. For gas systems, insulation is only applicable for high reservoir temperatures and/or short tieback lengths [17].

One advantage of an insulated production system is that it can allow higher water production, which would not be economical with continuous inhibitor injection. However, shutdown and restart operations would be more complicated. For example, long-term shutdowns will probably require depressurization [18].

6.6.10 HYDRATE REMEDIATION

Hydrate remediation is extremely difficult and very costly because of two factors, namely:

1. It takes large amount of energy to dissociate hydrates, and heat transfer through the hydrates is slow. Hydrate dissociation is a highly endothermic process; if the heat transfer through the pipeline insulation layer from the surrounding is low, then the temperature near the dissociating hydrate can rapidly reduce.
2. $1\,\text{ft}^3$ of methane hydrates can give up $160\,\text{ft}^3$ of methane gas. This has safe implications in depressurization. Increased JT expansion of the evolved gas can drastically reduce the surrounding temperature.

Therefore, either of the above mechanisms will form additional hydrates. Although the design of a unit is intended to prevent hydrate blockages, industry operators must include design and operational provisions for remediation of hydrate blockages. Some methods are listed in the Table 6.3.

TABLE 6.3
Hydrate Prevention/Remediation

| | Thermal/Thermodynamic | | Chemical | | |
	Passive System	Active System	For Hydrates	For Wax/ Paraffin	Mechanical
Prevention	Pipe-in-pipe	Electric heating	Dehydration	Dispersants	Pigging
	Buried pipe	Hot fluid circulation	Thermodynamic		
	External coated pipe	Exothermic process	inhibitors	Crystal modifiers	
	Synthetic jacketed pipe		Kinetic inhibitors		
Remediation		Electric heating	Dissociation	Solvents	Pigging
		Hot fluid circulation			
		Exothermic process		Microbes	Coiled tubing
	Depressurization				

Deep Sea Development

6.6.11 Selection of Hydrate Control Method

6.6.11.1 Gas System
- Gas systems are generally designed for continued injection of hydrate inhibitors because water production is small, typically only condensation water. Inhibitor requirements are, thus, relatively small, of the order of 1–2 bbl MeOH/mmscf.
- Pipe insulation is not preferred because of the possibility of JT effect.

6.6.11.2 Oil System
- As oil system produces free water, continuous inhibition would be a costly affair.
- For normal operations, insulation is preferred.
- For start-up and shutdown, a combination of methanol and blowdown (depressurization) is employed.

6.6.12 Wax Control Guidelines
- Operate above WAT (wax appearance temperature), and provide thermal insulation, if necessary.
- Operate at higher production rate to avoid deposition in wellbore and in tree.
- Pig frequently to ensure pig does not stick.
- Identify and treat high pour point oil continuously.

6.6.13 Wax Management Strategy

Once wax deposition is identified as a flow assurance issue to the production flow pathway, prevention and/or remediation techniques are investigated to determine their viability given the project's design and budget. The following wax management techniques are commonly examined:

- Prevention – Thermal control and chemical inhibition
- Remediation – Thermal and chemical wax dissolution and physical removal

6.6.13.1 Thermal Control
This system can assist in keeping the temperature above the cloud point for the entire flowline, thus eliminating wax deposition. Although these techniques are expensive, they can become financially attractive if other temperature-related flow assurance issues are to be mitigated as well, most notably hydrates.

6.6.13.2 Chemical Inhibition
Chemical inhibition is accomplished by deployment of wax inhibitors, also known as crystal modifiers. These chemicals are designed to affect the WAT and wax deposition propensity of the crude by disrupting the ordered aggregation of the

growing crystalline structure. Although complete wax inhibition is rarely achieved by chemical deployment, chemical usage can be a powerful synergistic component to an optimized wax management strategy.

Unlike wax solvents or dispersants, these chemicals must be deployed before the bulk temperature of the crude drops below the WAT, hence, they are often injected downhole.

6.6.13.3 Thermal and Chemical Wax Dissolution

This method uses the injection of heated fluid (e.g., crude oil, water) into the production pathway to melt and remove wax deposits. The fluid is heated well above the melting point of the wax and circulated to remove the deposition. Additives should also be included to the heated fluid to improve its application (e.g., wax dispersants, demulsifiers).

6.6.13.4 Physical Removal

Physical removal of wax deposits is accomplished in the wellbore and flowline by deploying wax/gauge cutters and pigs, respectively. There are multiple proprietary designs within the industry for both. To facilitate pigging, a dual-flowline system must be built with a design that permits pigging. Pigging must be carried out frequently to avoid the build-up of large quantities of wax. If the wax deposit becomes too thick, there will be insufficient pressure to push the pig through the line as wax accumulates in front of it. Pigging also requires that the subsea oil system be shut down, stabilized by a methanol injection and blowdown, and finally restarted after pigging has been completed [19] (Figure 6.45).

6.6.14 Asphaltene

Asphaltene problems occur infrequently in offshore, but can have serious consequences on project economics because asphaltene deposition is most likely to occur when the produced fluid passes through the bubble point. The deposition often occurs in the tubing.

FIGURE 6.45 Common pipeline pig [20].

Deep Sea Development

- Inject an asphaltene dispersant continuously into the wellbore (injection must be at the packer to be effective).
- Install equipment to facilitate periodic injection of an aromatic solvent into the wellbore for a solvent soak.
- Be financially and logistically prepared to intervene with coiled tubing in the wellbore to remove deposits.
- Control deposition in the flowline with periodic pigging using solvents.

6.6.15 CORROSION

The presence of carbon dioxide (CO_2), hydrogen sulfide (H_2S), and free water in the internal production fluid can cause severe corrosion problems in oil and gas pipelines. Internal corrosion in wells and pipelines is influenced by temperature, CO_2 and H_2S content, water chemistry, flow velocity, oil or water wetting, and the composition and surface condition of the steel.

Two types of corrosion can occur in oil and gas pipeline systems when CO_2 and H_2S are present in the hydrocarbon fluid: sweet corrosion and sour corrosion. Sweet corrosion occurs in systems containing only carbon dioxide or a trace of hydrogen sulfide (partial pressure <0.05 psi). Sour corrosion occurs in systems containing hydrogen sulfide above a partial pressure of 0.05 psia and carbon dioxide.

Normally during the design phase, corrosion margin based on crude property is provided to all the flow path material.

6.6.16 INTERNAL CORROSION PREVENTION

6.6.16.1 Internal Coating

An effective coating system provides an effective barrier against corrosion attack. The type of coating will depend on both environmental conditions and the service requirements of the line. The major generic coatings used for internal linings include epoxies, urethanes, and phenolic.

6.6.16.2 Internal Corrosion Inhibitors

Corrosion inhibitors are chemicals that can effectively reduce the corrosion rate of the metal exposed to the corrosive environment when added in small concentrations. They normally work by adsorbing themselves to form a film on the metallic surface. Inhibitors are normally distributed from a solution or by dispersion.

Inhibitors can be generally classified as follows:

- Passivating inhibitors
- Cathodic inhibitors
- Precipitation inhibitors
- Organic inhibitors
- Volatile corrosion inhibitors

6.6.17 External Corrosion Prevention

6.6.17.1 External Coating

Thick coatings are often applied to offshore pipelines to minimize the defects and resist damage by handling during transport and installation. Coatings must have good adhesion to the pipe surface to resist wear and tear.

The principal coatings, in rough order of cost, are:

- Tape wrap
- Asphalt
- Coal tar enamel
- Fusion-bonded epoxy (FBE)
- Cigarette wrap polyethylene (PE)
- Extruded thermoplastic PE and polypropylene (PP) [4]

6.6.18 Scales

Scales are deposits of different chemical composition because of the crystallization and precipitation of minerals from the produced water. These are generally inorganic salts such as carbonates and sulfates of metals calcium, strontium, and barium.

$$Ca^{2+} + CO_3^{2-} \rightarrow CaCO_3$$

$$Ca^{2+} + 2\left(HCO^-\right) \rightarrow CaCO_3 + CO_2 + H_2O$$

*Presence of CO_2 will increase the solubility of carbonates.

$$Ca^{2+} + SO_4^{2-} \rightarrow CaSO_4$$

$$Ba^{2+} + SO_4^{2-} \rightarrow BaSO_4$$

*$BaSO_4$ is extremely insoluble and almost impossible to remove chemically.

$$Sr^{2+} + SO_4^{2-} \rightarrow SrSO_4$$

*Strontium sulfate solubility increases with salinity.

6.6.19 Scale Management

6.6.19.1 Scale Inhibitors

These chemicals delay or prevent scale formation. Use of these chemicals is attractive because a very low dosage (several ppm) can be sufficient to prevent scale formation for extended periods. Common classes of scale inhibitors include:

- Inorganic polyphosphates
- Organic phosphates esters
- Organic phosphonates
- Organic polymers [4]

Deep Sea Development

6.6.20 Erosion

Erosion is a process of wearing out/loss of material surface due to abrasive/mechanical action resulting from fast flowing liquid or gas with suspended particles. Sometimes, it happens even due to the force of impinging liquid in the system.

6.6.21 Mitigation Methods

6.6.21.1 Reduction in Production Rate

Optimum production rate considering erosion, sand production, and economics should be selected [21].

6.6.21.2 Design of Pipe System

Blind tees are generally perceived as being less prone to erosion than elbows, hence, the use of fullbore valves and blind tees in place of elbows can reduce erosion problems. Minimizing the flow velocity and avoiding sudden changes (e.g., at elbow, constrictions, and valves) in the flow direction should be given much attention to reduce the severity of any erosion.

6.6.21.3 Increasing Wall Thickness

Thick-walled pipes are often used to increase the wear life of a pipe system.

6.6.21.4 Erosion-Resistant Material

If erosion problems are suspected, specialized erosion resistant materials such as tungsten carbide can be used. The primary factor of ductile materials in controlling erosion is their hardness. Consequently, steels are more resistant than other softer metals. In vulnerable components, specialized materials such as tungsten carbides, coatings, and ceramics are often used. These materials are generally hard and brittle and have a super erosion resistance to steel.

6.7 EMERGING DEEPWATER TECHNOLOGIES

Deepwater play starts in at water depth exceeding 500 m at the edge of continental shelf and beyond. The remote basins may have significant hydrocarbon deposits. According to the International Energy Agency, there could be approximately 270 billion barrels of recoverable oil alone in deepwater worldwide. Developing deepwater prospects is challenging and expensive owing to the harsh environments and complex reservoirs, which necessitates development of innovative technologies. Some of the emerging deepwater technologies are discussed below.

6.7.1 Dry Tree Semi-Submersibles

Dry tree floating platforms for deepwater offshore developments are typically spars and tension-leg platforms (TLPs). All production semi-submersibles in operation support only wet tree systems because the hull motions are not sufficiently small to allow the use of riser tensioning systems, which connect vertical production risers from subsea wells directly to the topsides. The wet tree semi-submersible designs can potentially be improved for dry tree applications.

Beyond 6,000 ft, TLPs become impractical because of the amount of steel needed for the tendons that moor the platform to the ocean bottom. Spars have virtually no water depth limit and are designed specifically as a dry tree unit. However, their size is limited by their cylindrical hulls, which constrain the available deck space, thereby pushing designers to stack the decks vertically. A conventional semi-submersible platform offers the optimum amount of deck space for safer operations and payload flexibility that a spar cannot, however, it has too much vertical motion for a dry tree to operate safely. Several semi-submersibles operate in depths exceeding 6,000 ft; however, all of them use subsea trees. Unlike subsea trees that are installed at the seabed and deliver hydrocarbons to a surface platform through a flexible production riser or a metallic riser such as a steel catenary riser (SCR), a dry tree uses a rigid riser system known as a top-tensioned riser (TTR) locked onto the subsea wellhead at the seafloor and to the dry tree at the platform deck, thereby making it very sensitive to movement. Although the DTS hull is designed to reduce motion caused by ocean forces, it is the motions that move the platform up and down that are of the highest concern. To compensate for excessive vertical motion current, DTS concepts are relying on proven riser tensioner technology [19] (Figures 6.46 and 6.47).

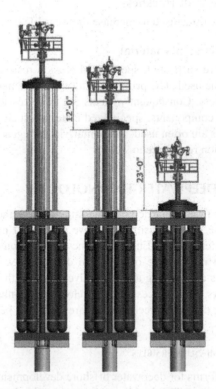

FIGURE 6.46 SBM Offshore's 35-ft range riser tensioner system allows its dry tree semi-submersible to move vertically while ensuring the integrity of the dry tree system located on the tensioner. (Source: SBM Offshore, JPT Sept. 2013.)

Deep Sea Development

FIGURE 6.47 The Octabuoy has a deeper draft than a conventional semi-submersible and can be deployed to multiple fields over its expected 50-year service life. The platform's unique column and octagonal buoyancy pontoon minimize the effects of heave, pitch, and roll [22].

Major Advantages of DTS:

- Minimizes flow assurance issues such as wax and hydrate formation
- Drilling, completion, and workover from single platform, eliminating the need of MODUs

6.7.2 Hybrid Riser System

The term "hybrid riser" generally refers to a system which incorporates both steel pipe and flexible pipe technologies. The steel pipe can be a vertical leg or an SCR connected to a subsurface buoy support and the flexible pipe is used to complete the fluid flow path from the subsurface buoy to the FPU. Hybrid riser arrangements have been conceived to meet the challenge of deep waters and minimize the drawbacks inherent to the steel riser systems, for example, coupling effect with FPU wave motions and subsequent fatigue issues faced by conventional risers. The hybrid riser systems provide reduced payload on the FPU and flexibility of field layout by permitting the routing of flowline to be independent of the approach lay azimuth angle of the riser top hang-off porch [23] (Figure 6.48).

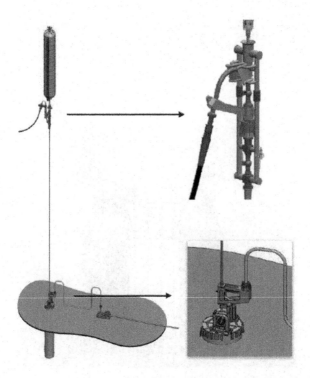

FIGURE 6.48 General configuration of hybrid risers [23].

Latest evolutions of hybrid riser systems:

- Free-standing flexible riser (FSFR)
- Multi-lines free standing riser
- Deep steep riser (DSR)

6.7.3 Free-Standing Flexible Riser System (FSFR)

The vertical leg of the riser system from seabed to the subsurface buoy is a flexible pipe instead of a rigid steel pipe. The FSFR is a concept developed to optimize the riser installation tension and in-situ dynamic behaviors (Figure 6.49).

Benefits include:

- Minimization of total loads applied to FPU
- Decoupling between riser installation and FPU arrival
- Eliminates the requirement of offshore heavy lifting

6.7.4 Multi-Lines Free Standing Riser

Multi-lines free standing riser system is an innovative concept, gathering several rigid risers to be fully assembled offshore and hung onto a single top riser assembly/buoyancy tank system.

Deep Sea Development 197

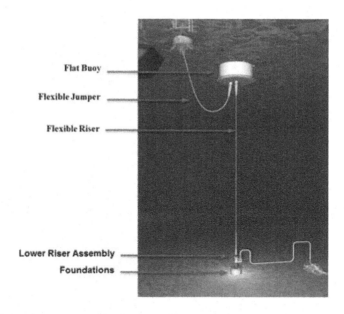

FIGURE 6.49 Free-standing flexible riser [23].

These are developed to provide the following functional requirements:

- Production line combined service line to form an "hybrid loop" for round-trip pigging.
- Water injection line to maintain reservoir pressure and production.
- Gas lift at the production riser base for produced fluid artificial lift (Figure 6.50).

6.7.5 Deep Steep Riser

This innovative concept consists of a single leg tensioned riser, which can be either "full flexible" or "hybrid" riser solutions. The riser system is basically composed of a flexible jumper at its upper section to decouple from the FPU motions. The lower riser vertical section can be either a flexible or a rigid steel riser to reach the seabed foundation. Applicable for ultra-deepwater developments (4,000 m+) (Figures 6.51 and 6.52).

6.7.6 Expandable Monobore Liner Extension

The goal of an expandable monobore is to enable the operator to optimize drilling casing programs by drilling deeper wells with larger hole sizes to the reservoir. As a contingency plan, the goal is to enable the operator to isolate zones that contain reactive shales, subsalt environments, low-fracture-gradient formations, or other drilling situations. Utilization of this technology will allow wellbore construction to continue without having to reduce the casing and subsequent drilled hole size. Expandable liner system employs a top-down expansion system to extend a casing string while maintaining the same casing drift. The system provides an optimized, cost-effective

FIGURE 6.50 Rigid risers hung to top riser assembly in multiline free standing riser [23].

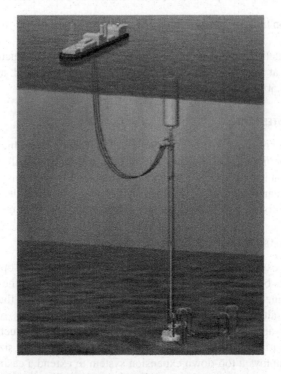

FIGURE 6.51 Multi-line free standing riser [23].

Deep Sea Development

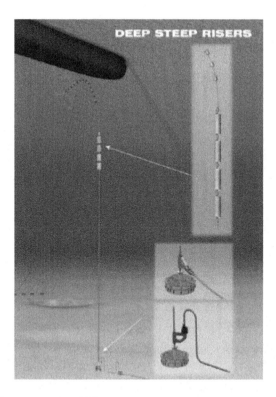

FIGURE 6.52 Deep steep riser [23].

casing configuration without the reduced hole size encountered in typical telescoping casing programs or with other solid expandable drilling liners (Figure 6.53).

Installation of the first monobore expandable liner extension system (MELES) has proven the feasibility of solid expandable tubulars as an alternative to conventional, telescoping casing designs. MELES could enable drilling deeper wells with larger hole sizes at the reservoir, and isolating zones with reactive shales, subsalt environments, low-fracture-gradient formations, or other such situations, and then drill ahead without having to reduce the casing and subsequent drilled hole size.

6.7.7 Smart Well Technology

A smart well completion system optimizes production by collecting, transmitting, and analyzing completion, production, and reservoir data, allowing remote selective zonal control and ultimately maximizing reservoir efficiency by:

- Helping reduce capital expenditure – The ability to produce from multiple reservoirs through a single wellbore reduces the number of wells required for field development, thereby lowering drilling and completion costs. Size and complexity of surface handling facilities are reduced by managing water through remote zonal control.

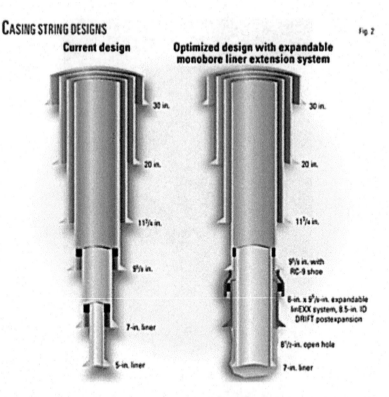

FIGURE 6.53 Comparison of convention telescopic completion and expandable monobore liner extension system. (Source: lin EXX – Baker Hughes.)

- Helping reduce operating expenditure – Remote configuration of wells optimizes production without costly well intervention. In addition, commingling of production from different reservoir zones shortens field life, thereby reducing operating expenditures.

Smart wells include intelligent completions, with remote ICVs, flow control, measurement sensors (flow, pressure, and temperature). The data can be conveyed wirelessly using satellite communication to the base station [24] (Figure 6.54).

6.7.8 Autonomous Underwater Vehicles (AUVs)

AUVs are programmable, robotic vehicles that, depending on their design, can drift, drive, or glide through the ocean without real-time control by human operators. Some AUVs communicate with operators periodically or continuously through satellite signals or underwater acoustic beacons to permit some level of control. Compared to remotely operated vehicles (ROVs), AUVs operate without an umbilical and are able to move faster and more quietly, collecting data at a very high data-to-signal ratio. As a result, surveys are completed sooner and with more accuracy (Figure 6.55).

Deep Sea Development

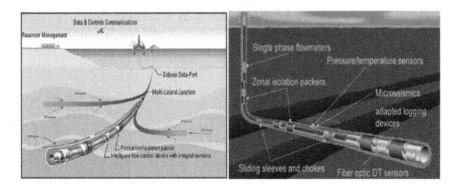

FIGURE 6.54 Smart wells with intelligent completions [25].

FIGURE 6.55 Autonomous underwater vehicle (AUVs) [26].

Benefits of autonomous inspection:

- Fewer personnel at sea
- Large stand-off distances from the facility being inspected
- No divers or ROV umbilical
- Underwater homing and docking of the marlin
- Controlled launch and recovery
- Reduced weight
- Smaller footprint [26]

6.7.9 Nomad Systems

The nomad system consists of a small floating unit equipped with minimum facilities to produce subsea wells and to pump the produced effluent in a single multiphase pipeline, towards an existing platform, for separation, processing, and export. It is designed to operate in an unmanned mode.

The floating unit is anchored on the seabed by a catenaries' mooring system. Its deck supports the pumping and power generation systems. It also contains the control and safety systems as well as a survival shelter.

- Lowest development cost
- Lowest OPEX
- Unmanned operations
- Logistics via supply boats
- Self-supporting (wind and solar energy)

6.7.10 Subsea Multiphase Pumps

Multiphase production systems require the transportation of a mixture of oil, water, and gas, often for many miles from the producing well to a distant processing facility. This represents a significant departure from conventional production operations in which fluids are separated before being pumped and compressed through separate pipelines. By introducing this concept the cost of pumping facility is approximately 70% that of a conventional facility and significantly more savings can be realized if the need for an offshore structure is eliminated altogether (Figure 6.56).

However, multiphase pumps do operate less efficiently (30%–50%, depending on gas volume fraction and other factors) than conventional pumps (60%–70%) and compressors (70%–90%).

6.7.11 Subsea Processing

Subsea processing consists of a range of technologies to allow production from offshore wells without needing surface production facilities. Subsea processing using subsea separation and pumping technologies has the potential to revolutionize

FIGURE 6.56 Multiphase pump. (Source: FMC Technologies.)

Deep Sea Development

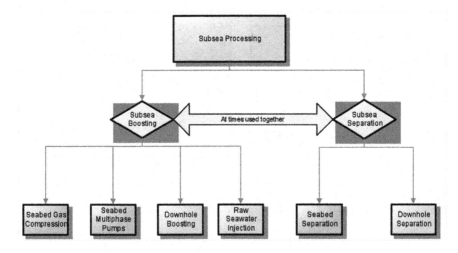

FIGURE 6.57 Typical subsea processing flow diagram.

offshore oil and gas production. It can reduce development cost, enhance reservoir productivity, and improve subsea system reliability and operability when combined with relatively mature subsea production technologies (Figure 6.57).

The benefits of introducing subsea processing in a field development include:

- Increase ultimate recovery by extending economic life.
- Improve well productivity with greater pressure drawdown.
- Avoid gas hydrates with no inhibition or with reduced inhibitor dosage.
- Allow online pigging to control wax deposition in oil pipelines.
- Eliminate fluid surges by use of single-phase pipelines.
- Prevent solids dropout by allowing higher liquid-flow velocities.
- Reduce capital and operating costs by reducing surface processing needs [27].

The two technologies discussed in detail in this analysis are seabed separation and seabed boosting.

6.7.12 SEABED SEPARATION

Oil, gas, and water separation is essential for further processing of well products. Seabed separation of the product is being restored with increasing numbers of deep-sea field. Water production starts increasing in mature fields with corresponding decline in oil production, resulting in increased cost of production from a declining reservoir. In such a case, separation of oil, gas, and water at seabed becomes economical. This technology can be used in green field technology also starting from initial production on the basis of long-term production profile or in fields with high risk of hydrate formation.

The key parameters specifying either oil/water or liquids/gas separation includes increased water depths and a number of fields tied back to a platform facility. The key factors involved in oil and water separation in green and brown fields are obviously

FIGURE 6.58 (a) Oil/water separator (Tordis in North Sea), (b) Gas/liquid separator (Pazflor in Gulf of Mexico).

not similar. In mature/brown fields, the level of the field's water production and the existence of heavy oil are key factors, whereas high gas volume fraction, increased distance from the host, and low reservoir pressure and temperature are key factors in green fields [28].

Some examples of seabed separators are shown in Figure 6.58.

6.7.13 Subsea Pressure Boosting

The aim of pressure boosting is to increase the flow pressure in the production system when the reservoir pressure is insufficient to transport produced fluids at an optimum rate. Seabed pressure boosting is at times deployed to ensure fluid flow from fields at the required rate after natural reservoir pressure declines. Some key parameters that lead operators to use seabed booster pumps include the existence of heavy oil, increased distance from the host, increased water depth, low reservoir pressure, and a greater number of fields tied back to the host [29].

Another technology that has established itself recently in multi-phase production is the electrical submersible pump (ESP) used on the seabed instead of downhole. Pressure boosting may be necessary at any given stage of the project development; therefore, it is an important consideration in the design stage of the project. Finally, seabed gas compression involves gas compression at the seabed level instead of gas compression on a topside facility. Key factors driving the implementation of subsea gas compression technology are the discovery of distant offshore gas fields, increased water depths, long step-outs from the host facility, harsh environmental conditions, and low reservoir pressure and temperature.

Components of a subsea boosting station may include:

- *Subsea gas compressor*: used for gas re-injection into the reservoir for pressure maintenance;
- *Subsea multiphase pump*: used to reduce the back-pressure on wellheads, thus increasing the transport distance;
- *Subsea wet gas compressors*: used for gas transportation to remote offshore host facilities or onshore factory (Figure 6.59).

Deep Sea Development

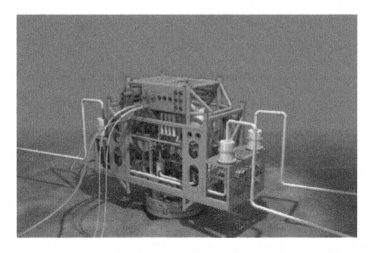

FIGURE 6.59 Subsea pressure boosting pump [29].

REFERENCES

1. B. Oyeneyin, *Introduction to deepwater field development strategies*, Dev. Pet. Sci. 63 (2015), pp. 1–9.
2. *ONGC, in Graduates's Guide to Offshore Operations*, 2005.
3. *Subsea_(technology)*. Available at https://en.wikipedia.org/wiki/Subsea_(technology).
4. Y. Bai and Q. Bai, *Subsea Engineering Handbook*, Gulf Professional Publishing, Houston, TX (2012).
5. E.A. Ageh, O.J. Uzoh, and I. Ituah, *Production Technology Challenges in Deepwater Subsea Tie-Back Developments* (2011).
6. *Subsea Umbilical Termination Assembly (SUTA)*. Available at www.oilfieldwiki.com/wiki/Subsea_Umbilical_Termination_Assembly_(SUTA)
7. *Hydraulic_Power_System*. Available at http://oilfieldwiki.com/wiki/Hydraulic_Power_System
8. *"Oceaneering,"*. Available at www.oceaneering.com
9. *Subsea_distribution_system*. Available at www.oilfieldwiki.com/wiki/Subsea_distribution_system
10. C.G.J. Nmegbu and L.V. Ohazuruike, Subsea technology : A wholistic view on existing technologies and operations, *Int. J. Appl. Innov. Eng. Manage.* 3 (2014), pp. 81–87.
11. *subsea-technology-and-equipments*. Available at www.oil-gasportal.com/subsea-technology-and-equipments/
12. *OilField Wiki*. Available at www.oilfieldwiki.com
13. *Hydroacousticsinc*. Available at www.hydroacousticsinc.com
14. *Wax_Management*. Available at www.oilfieldwiki.com/wiki/Wax_Management
15. *Sbsea_Accumulator_Module_(SAM)*. Available at www.oilfieldwiki.com/wiki/Subsea_Accumulator_Module_(SAM)
16. *Travaux des élèves*. Available at http://hmf.enseeiht.fr/travaux/bei/beiep/book/export/html/1772
17. D. Denney, Active heating for flow-assurance control in deepwater flowlines, *J. Pet. Tech.* 56 (2004), pp. 45–46.
18. R.G.S. Cochran, *Hydrate Management: Its Importance to Deepwater Gas Development Success*, World Oil.

19. *Wax Management Startegy.* Available at www.gateinc.com/gatekeeper/gat2004-gkp-2016-10
20. *Apache Pipe.* Available at www.apachepipe.com/
21. E.S. Venkatesh, *Erosion Damage in Oil and Gas Wells* (2007).
22. T. Jacobs, Dry tree semisubmersibles: The next deepwater option, *J. Pet. Tech.* 65 (2013), pp. 1–6.
23. A. Luppi, G. Cousin, and R. O'Sullivan, *OTC 24802-MS Deep Water Hybrid Riser Systems*, (2014), pp. 1–13.
24. G.A. Carvajal, I. Boisvert, and S. Knabe, *A Smart Flow for SmartWells: Reactive and Proactive Modes.* Society of Petroleum Engineers, Utrecht (2014).
25. *Oil-gasportal.* Available at www.Oil-gasportal.com
26. D. Mcleod and J. Jacobson, *Advances in Autonomous Deepwater Inspection* (2014).
27. O.T. McClimans and R. Fantoft, *Status and New Developments in Subsea Processing*, (2008), pp. 1–10.
28. R. Perry, *Advances in Subsea Recovery*, Offshore (2012).
29. *Subsea Boosting Offers Field development Solutions.* Available at www.hartenergy.com/exclusives/subsea-boosting-offers-field-development-solutions-176414

7 Offshore Field Development

7.1 INTRODUCTION

Different aspects of offshore field operations and development have been discussed in respective chapters.

1. Normal sequence of field development is collection, collation, and interpretation of primary and secondary geological and geophysical parameters.
2. Preliminary assessment and evaluation of field potential based on point 1 above.
3. Acquisition, collation, and interpretation of additional geological and geophysical data to support strategy based on preliminary assessment and evaluation as per point no 2 above for development.
4. Basic concept of development and primary technoeconomical evaluation based on the abovementioned points.
5. Concept selection and basic engineering.
6. Detailed engineering.
7. Execution and commissioning.

To demonstrate the different aspects of field development, two examples are discussed below for better understanding, namely, one for marginal field and another for giant field.

7.2 OFFSHORE MARGINAL FIELD DEVELOPMENT EXERCISE

There are several approaches to developing an offshore oil field. The best development plan is one that ensures viability throughout the life of the field. Adequate planning is important because we need to commit to a production system right from the early developmental stages of the field when relatively little is known about the field as well as its reserves.

7.2.1 DEVELOPMENT EXAMPLE

Offshore marginal fields need to be developed as per the following parameters:

Water depth: 120 m
Distance from shore: 160 km
Production rate: 20,000 bopd (max)
Production per well: 2,500 bopd (max)

Production wells: 6
Injection wells: Nil
GOR: 560 scf/bbl
Sulfur content: 230 ppm
Carbon dioxide content: 3%
Field life: 5 years (approx.)

7.2.1.1 Field Development Scenarios: Options/Alternatives

Two probable field development scenarios based on the utilization of floating production systems for the field include the following:

Option A: FPSO and subsea well development
Option B: Semi-submersible FPS and subsea template well development

7.2.1.1.1 Option A

FPSO and subsea well development: We may have to fully understand the components of field development, as well as the field's operation, storage, offloading (of crude oil), as well as mooring and anchoring requirements in the overall subsea production system.

7.2.1.1.1.1 Field Development Components
- FPSO
- Subsea wells
- Flexible flow lines, annulus lines, and control umbilical connecting the wells to the tanker
- Shuttle tankers

7.2.1.1.1.2 Understanding Major Components and Operations
- FPSO
- Process facilities
- Emergency shutdown systems (ESD)
- Firefighting and safety systems
- Other utility systems
- Crude oil storage and offloading
- **FPSO** FPSO are required to store production for a period of time equal to a round trip of a shuttle tanker (5.5 days).
- **Mooring system and tanker motion**
  A single point mooring system is selected so that the tanker can take up positions with the least resistance from wind, waves, and currents. In the given field and sea environment, SBM's turret mooring column system equipped with the quick disconnect feature was selected as it is the best option.
- **Subsea production system**
  Because the selected tanker mooring system could support the required number of production risers, a subsea manifold is not proposed here.

The completion system is a conventional, guideline, subsea wellhead system enabling remote installation and testing without the aid of divers.

Offshore Field Development

The tree assembly is a wet tree assembly providing control for production, well control, annulus monitoring, flushing of all lines, and pigging capability.

7.2.1.1.2 Option B

Semi-submersible FPS and subsea template wells:

- Field development components
- Semi-submersible vessel with onboard process facilities
- Subsea well template with an integral subsea manifold
- Flexible flow lines/risers connecting the subsea manifold to the FPS
- CALM loading buoy and temporary storage/shuttle tankers

There are definite advantages of semi-submersible-based FPS compared to tanker-based FPS; when the field is abandoned, the rig can recover the subsea trees and downhole completion, thus significantly reducing the field's abandonment costs.

Although the initial cost is high, they are offset against the system's flexibility for reuse over a wide range of water depth.

The semi-submersible FPS, subsea template and wells, and flexible riser systems present no technical concerns, as they have been successfully used in several similar applications elsewhere.

7.2.2 OFFSHORE GIANT FIELD DEVELOPMENT EXERCISE

The field under study is a giant offshore multilayered reservoir separated by impermeable 1–2 m thick shale layers, and appraisal drilling is almost complete. The required geological, logging, core, PVT, fluid analysis, production testing, and pressure transient test data have been generated. Prima facie, the field appears to be operating under partial edge water drive mechanism.

7.2.2.1 The Salient Data Is

Water depth: 80–100 m
Average reservoir depth, msl: 1,400 m
Initial reservoir pressure: 160 kg/cm^2
Initial reservoir temperature: 120°C
Average porosity: 0.20 (range 0.10–0.35)
Permeability: 50–3,000 mD
API gravity: 38° (Sweet crude)
Bubble point pressure: 160 kg/cm^2
μ_o at res temp: 0.45 cP
B_{oi}: 1.35 v/v
$R_{si} = 110 \, \text{m}^3/\text{m}^3$
Resin = 8–12%
Wax content = 15–20%
Pour point = 36–39°C
Salinity of formation water: 23 g/l
PI = 20–40 m^3/day/kg/cm^2

Based on the above information, work out the field development plans considering the following:

1. Drilling rig type and completion strategy
2. Water injection requirement, if any
3. Artificial lift selection
4. Type of process platform, processing, and transportation facilities

7.2.2.2 Solution Approach

1. Drilling rig type and completion strategy (Figures 7.1–7.3)
2. Water injection requirement, if any

As the field is operating under partial edge water drive, peripheral water injection scheme shall be required as a void age compensation technique for pressure maintenance and improved recovery (Figures 7.4 and 7.5).

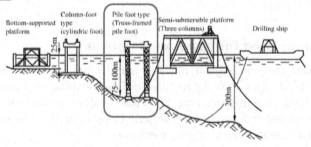

FIGURE 7.1 Comparison of different offshore drilling rigs for different water depths. (source: www.books.google.co.in.)

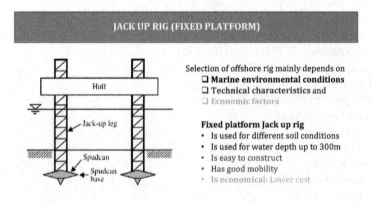

FIGURE 7.2 Spudcan foundation and drilling rig source (Lee and Randolph, 2011).

DEVELOPMENT & COMPLETION STRATEGY

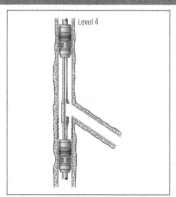

TAML: Technology Advancements – Multi Laterals

TAML level 4:

Main borehole: Cased & Cemented
Lateral borehole: Cased and cemented at junction

Excellent Pressure integrity at junction pointas main borehole and lateral section cased and cemented

FIGURE 7.3 TAML level 4 completion configuration. (Source: www.slb.com)

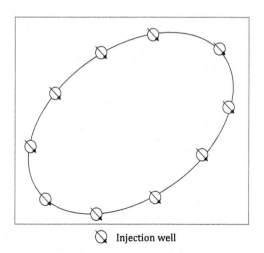

FIGURE 7.4 Schematic of peripheral water injection pattern (top view).

3. Artificial lift selection

   Available data suggests continuous gas lift to be the most suitable artificial lift technique (Figure 7.6).
4. Type of process platforms, processing, and transportation facilities

   Being a giant field, it should be developed with an integrated approach at the grassroots level:
   1. Unmanned well platforms (solar-powered) with testing facilities.
   2. Centrally located jacketed process platform with required utilities including living quarter. Only one stage/first stage separation to be done. Separated gas to be used for power generation and Gas lift. Oil and gas

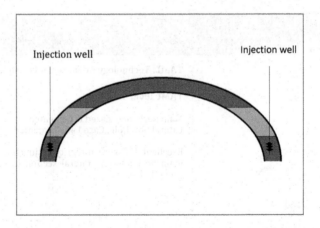

FIGURE 7.5 Peripheral water injection pattern (cross-sectional view).

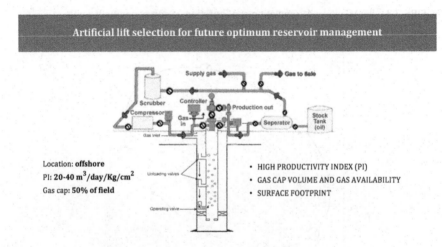

FIGURE 7.6 Schematic of gas lift surface and subsurface facility.

to be pumped using MOL pumps for further onshore processing. Until a pipeline is available, a shuttle tanker to be used to transport oil to onshore terminals.
3. Water injection and gas compression facilities to be located on separate platforms and connected with a bridge to the process platform to share utilities and other requirements.

Entire operation to be controlled and monitored using SCADA.

8 Health, Safety, and Environment in Offshore

In petroleum industry, the probability of risk is high. An accident in the industry, particularly in offshore, can turn into a disaster, resulting in the loss of life and property, as well as damage to the environment. Human resources have to be evolved and upgraded with the view to minimizing the overall risk to human life rather than containing damage to property and environment as the value of life is incomparable. This is being achieved by meaningful training [1] focusing on awareness and alertness, which is the key to minimize damage to property and environment.

8.1 BASIC DEFINITIONS

1. Safety: Control of accidental loss or hazards is known as safety. It involves constant awareness to critical work hazards through a constantly improving system [1].
2. Hazard: The potential to cause harm to people, property, or environment is known as hazard. Any loss of life, profits, business, reputation, skills, etc. is known as hazard effect [1].
3. Risk: Risk is a combination of the hazard effect and the probability that harm to people, property, or environment will actually occur [1].
   • Risk = Hazard effect × Probability of occurrence [1].
4. Accidents: Accidents are undesirable events that result in harm to human beings and damage to property and environment [1].

8.2 HUMAN FACTORS

• The oil and gas industry has a major accident potential. Both exploration and production rely on advanced human–machine interfaces, and include activities with a complex organizational structure. Increasingly, the work is performed by distributed teams and remotely controlled technology. Human factors have become an important and integral part of the industry's approach to safe and efficient operations [2].
• Human factors refer to the application of what we know about human capabilities and limitations to maximize overall system performance. By giving careful consideration to the interactions between humans and technological and organizational elements of a system, it is possible to significantly increase the system's productivity and reliability [2].
• Human factors address the interaction of people with other people, with facilities, and with management systems in the workplace. These factors have been shown to have an impact on human performance and safe

213

operations. They provide practical solutions to reduce incidents while improving productivity [2].

- In the oil and gas industry, human factors are an essential component in the effort to operate in a safe and efficient manner. Areas where human factors have a key role include [2]:
  - Design of tools, equipment, and user interfaces in a way that augments the user's work performance
  - Human and organizational factors in risk assessment and emergency preparedness planning
  - Human behavior and cognition in accident causation
  - Efficient decision making and teamwork in stressful or critical situations
  - Safety culture and safety behavior improvement programs
  - Organizational reliability
- Human factors aim to achieve outstanding performance by proactively identifying risks and improvement opportunities, promoting safety leadership and designing improvement strategies, applying best practice tools, and supporting implementation to business and operational functions [2].
- To achieve efficient and safe operations, the following human factor effectiveness (HFE) is considered:
  - Establish management commitment to HFE and appoint an HFE Champion (management commitment is essential; the HFE Champion promotes HFE issues and provides a link between the HFE practitioner and the project management team).
  - Apply HFE during all phases of a project.
  - Provide an early focus on known HFE problem areas and lessons learned from other facilities.
  - Locate HFE personnel within the engineering design team to foster discussions and trust. It is more effective to act as an integrated part of the design team rather than an external enforcer of standards.
  - Mandate HFE in the project design and mandate accepted HFE design standards in project specifications.
  - Incorporate HFE activities into the program plan.
  - Provide management oversight of HFE activities.
  - Require close cooperation between HFE, operations/maintenance, and other engineering disciplines throughout the project lifecycle [3].
  - Engage academically educated and experienced HFE professionals. HFE is a unique engineering discipline that applies specific knowledge of human capabilities and limitations and should not be assigned to other engineers from other disciplines, health professionals, or former operators/maintenance staff.
- As a relatively new discipline, it is important to effectively manage the integration of HFE on projects. Utilizing suitably qualified personnel to apply the appropriate tasks in a suitable organizational climate should

Offshore Health, Safety, and Environment 215

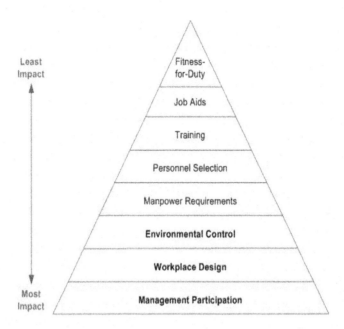

FIGURE 8.1 The human and organizational factors "Triangle of Effectiveness" model for reduction of human error [4].

maximize the success and impact of HFE on projects. Figure 8.1 displays the impact caused by various human and organizational factors to reduce human errors.
- In addition to human factors, a proper lifetime assessment of offshore structures is required, which is dealt briefly in this chapter in the section Life Extension and Assessment.

8.3 HAZARDS ON OIL AND GAS INSTALLATIONS

Various hazardous incidents may occur on offshore process installations. The following are the major hazard possibilities for offshore platforms with their detailed description [1].

1. Hydrocarbon release:
   - Blowout
   - Loss of containment or hydrocarbon release due to leakage or equipment or pipeline failures
2. Collision events:
   - Passing/drifting vessel collision
   - Supply boat collision
   - Helicopter crash
   - Dropped object

216　Offshore Operations and Engineering

3. Structural collapse:
   - Earthquake
   - Extreme wind and waves
   - Member failure due to any reason

8.4 PROCEDURAL ASPECTS RELATED TO SAFETY

It is essential to develop system, methods, and procedures to perform safe operations. As an outline, few aspects are mentioned below which needs to be developed or written by the operating company/personnel:

1. Standard operating procedures
2. Standard maintenance procedures
3. Safety management system
4. Operations-specific safety standards/charts
5. Risk/Hazard analysis
6. Work permit system
7. Safe system for working on an electrical system (including lockout and tagout system)
8. Management of change system

8.4.1 SYSTEM SAFETY

8.4.1.1 Process Safety and Hydrocarbon Release

Process safety and related issues have been discussed in Chapter 4 (Dealing with Production). Some points are illustrated below regarding the process and hydrocarbon leak.

8.4.2 PROCESS LEAKS

- The process event includes either gas, well fluids, or oil leaks from pigging facilities, well fluid heating, separation, gas compression, and gas dehydration equipment [1].
- To prevent a hazard, leaks should be isolated from ignition sources [1].

8.4.3 RISER LEAKS

Risers are a vital part of an offshore oil installation. The threat to installation from risers is of utmost concern as the means of inventory isolation in such cases are limited. The leaks through high-pressure gas riser have a relatively high potential for major hazard risks. Possible alternative protection measures and remedial actions for minimization of riser hazard include [1]:

- Relocation of the riser to increase its separation from other risers.
- Fire proofing/coating of the riser to increase protection in the event of flame impingement.

- Subsea isolation valves.
- Upgradation of the fire resistance capability of the firewalls.

8.4.4 Fire and Gas Detection and Safety System

8.4.4.1 Fire Protection

Fire protection is the study and practice of mitigating the unwanted effects of potentially destructive fires. It involves the study of the behavior, compartmentalization, suppression, and investigation of fire and its related emergencies, as well as the research and development, production, testing, and application of mitigating systems.

Protection from fire requires knowledge of fire, fire safety alarm system, and fire safety equipment. Knowledge of fire gives us an idea about the type of fire and the type of fire extinguisher to be used. Fire safety alarm system and fire safety equipment are used to detect and prevent fire [1].

Fire protection is of two types:

- **Passive fire protection**: The installation of firewalls and fire-rated floor assemblies to form fire compartments intended to limit the spread of fire, high temperatures, and smoke
- **Active fire protection**: Manual and automatic detection and suppression of fires, such as fire sprinkler systems and fire alarm systems.

8.4.4.2 Fire

Fire is the rapid oxidation of a material in the exothermic chemical process of combustion, releasing heat, light, and various reaction products. Figure 8.2 describes the fire tetrahedron.

The fire tetrahedron is a four-sided geometric representation of the four factors necessary for fire: fuel (any substance that can undergo combustion), heat

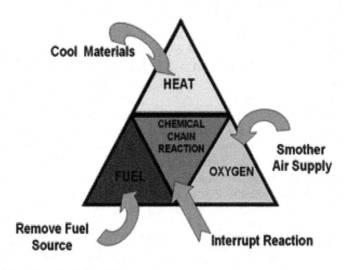

FIGURE 8.2 Fire tetrahedron [5].

(heat energy sufficient to release vapor from the fuel and cause ignition), oxidizing agent (air containing oxygen), and uninhibited chemical chain reaction (sufficient exothermic reaction energy to produce ignition). The fuel/air ratio must be within flammable limits, which describes the amount of vapor in air necessary to propagate the flame. Removing any of these four factors will prevent, suppress, or control fire.

8.4.4.2.1 Classification of Fire

- Class A: Fires involving ordinary combustible materials, such as wood, cloth, paper, rubber, and many plastics.
- Class B: Fires involving flammable liquids, greases, and gases.
- Class C: Fires involving energized electrical equipment.
- Class D: Fires involving combustible metals, such as magnesium, titanium, zirconium, sodium, and potassium.
- Class K: Fuels in this class are similar to Class B fuels, but involve high-temperature cooking oils and therefore have special characteristics. Class K agents are usually wet chemicals, water-based solutions of potassium carbonate-based chemicals, potassium acetate-based chemicals, or potassium citrate-based chemicals, or a combination.

8.4.4.2.2 Fire Detection Systems

1. UV Detection System

   UV detectors are utilized to give the earliest possible detection of rapidly developing gas and oil fires. Detectors are sensitive to radiation over the range of 1,850–2,450 Å and insensitive to light. UV detectors are not affected by wind, rain, humidity, or extremes of temperature or pressure and are suitable for both indoors and outdoors [1].

2. Thermal and Smoke Detection System

   Thermal detection systems detect an increase in temperature caused by a fire, initiate an alarm signal, and actuate the automatic extinguishing system for that particular area. When thermal detectors are located in an enclosed room protected with a Halon 1,301 system, the signal is transmitted to the fire and gas panel, which in turn produces audible and visible alarms, sends a shutdown signal to the HVAC unit, and actuates the Halon system [1].

3. Gas Detection System

   A gas detection system is provided to detect flammable and combustible gases before they reach a concentration that would cause a fire or explosion. They are the first line of defense in prevention of human injury and equipment damage. The layout of gas detectors on offshore facilities is determined by two philosophies [1]:

   1. Gas detectors shall be located where leaks are most likely to occur.
   2. Gas detectors shall also be located where the consequences of gas accumulation is the greatest.

   Gas detectors are calibrated for methane and hydrogen as the gases. Gas detectors calibrated for hydrogen only are installed in the battery rooms.

4. Fusible Plug Loops

Firewater header is always maintained at a pressure of 10 KSC, and all open deck areas in the complex are covered with a spray system. Fusible plug loop is a ring of pressurized air tubing, which contains fusible plugs at fixed intervals. This loop covers all the open deck process areas. When these plugs are in contact with heat, they melt down, and the pressure of this loop is vented to the atmosphere [1].

8.4.4.2.3 Fire Suppression Systems

1. Fire Water System:

The fire water system primarily consists of equipment to pump and distribute water for firefighting purposes. The water for this system is salt water taken from the sea. The pump takes suction from the sea and starts automatically upon the command of FSD from the fire and gas panel. Fire water sprinkler/deluge loops (Figure 8.3) located all over the open areas spray water over the equipment and vessels [1].

2. Spray System (Figure 8.4):

Water is the most effective firefighting medium. The main use is to provide cooling, control the fire, and reduce the risk of explosion. The combustible and flammable liquids are cooled to below their flashpoint, and hence the combustion cannot be sustained. A spray deluge system consists of a deluge valve, open-type spray nozzles, fusible plugs, and fire shutdown valves. The spray system is actuated automatically or manually [1].

3. Sprinkler System

Sprinkler systems are installed to extinguish Class A fires, which are likely to occur in living quarters, workshops, and storerooms. The sprinkler system includes manually operated isolation valves, which normally are locked open. If a fire occurs in a room of the living quarters, the smoke detector will be actuated by combustion gas and particles, and fusible metal of the sprinkler head will be melted by the heat generated by fire [1].

FIGURE 8.3 Spray system [1].

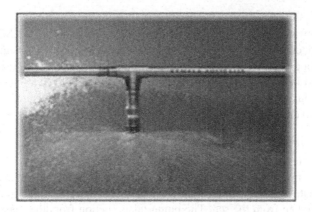

FIGURE 8.4 Spray system [1].

FIGURE 8.5 Firefighting vessels and fire extinguishers [1].

4. Dry Chemical Fire Fighting System:
   Dry chemical powder is important because it is the most powerful three-dimensional medium available. Powder is effective as an airborne cloud of small particles, which inhibits the chemical reaction of fire. Dry chemical extinguishing systems are considered satisfactory protection for flammable or combustible liquids, combustible gases, and electrical hazards [1].
5. Other fire suppression system:
   Halon system, firefighting vessels, fire extinguishers, etc. [1] (Figure 8.5). Now a days, due to environmental concerns, any Ozone depleting sustenance is not used as fire suppression agent.

8.4.5 Safety in Logistics Operations Related to Offshore Installation

Logistic operations in offshore are very risky and hazardous, involving transfer of man and material from sea or air to platform and vice versa. These operations involve the risk of collision, drowning in sea, dropping of object, etc.

FIGURE 8.6 Showing offshore production platform before and after collision with the ship (left: previous and right: after) [6].

8.4.5.1 Vessel Collisions

Normally, offshore facilities are installed away from sea traffic route to minimize any serious incident/accident with big ships. Nonetheless, the chances of collision with isolated drifting ship/vessel cannot be ruled out. Every company has a contingency plan for avoiding or mitigating such incidents or otherwise coastal state provides support in such an eventuality. Nowadays, with the availability of satellite communication, the level of preparedness has increased.

At the same time, collision/touching or hitting of platform by supply boat or support boat or loading tanker cannot be ruled out. Every company has their own approved standard operating procedures to avoid and minimize such incidents. Loading tankers are normally assisted by tugboats of adequate capacity to avoid such occurrences while loading crude from offshore platform; nonetheless, it has become more important with the advent of FPSO [1]. These collisions sometimes can damage the entire platform and can cause heavy economic losses, as shown in Figure 8.6.

8.4.5.2 Helicopter Incidents

Transportation of crew members and amenities is very difficult in offshore. A helicopter is used to transfer manpower from the shore to the offshore. Helipad of adequate size and approved type is normally provided on offshore facilities/installations to receive helicopters in favorable weather conditions. Offshore-going helicopters are normally fitted with flotation mechanism to avoid immediate submergence in water in case of crashing in sea.

In addition, it has been made mandatory for every personnel working on offshore platcorm or visting an offshore platform to undergo helicopter underwater escape training (HUET) (Figure 8.7).

8.4.5.3 Dropped Object

Involuntary dropping of any object in sea or deck during material handling may have very severe consequences. It may lead even to fatalities in addition to damage to above water or under water assets. Normally, handling of heavy objects by crane

FIGURE 8.7 Recovered crashed helicopter [1].

or other lifting equipment is done in safe areas to minimize any damage. Moreover, most of the assets, either under water or above water, are designed with dropped object protection to a greater degree.

Dropped objects can be can be caused by some of the following events:

1. Failure associated with lifting equipment
2. Load mismatch
3. Human failure related to handling
4. Unfavorable weather condition

8.4.6 EVACUATION, ESCAPE, AND RESCUE (EER)

Evacuation, escape, and rescue (EER) is the immediate evacuation from a place or escape of people away from an area that contains an imminent threat, an ongoing threat, or a hazard to lives or property. Examples range from small-scale evacuation of a building due to a storm or fire to the large-scale evacuation because of a storm, accident, or approaching weather system. In situations involving hazardous materials or possible contamination, evacuees may be decontaminated prior to being transported out of the contaminated area [1].

8.4.6.1 Reasons for EER
- Natural disasters
- Fires
- Accidents

8.4.6.2 Evacuation Sequence

The sequence of an evacuation can be divided into the following phases:

- Detection, decision, alarm, reaction, movement to an area of refuge or an assembly station, and transportation.
- The time for the first four phases is usually called premovement time.
  - The particular phases are different for different objects, for example, for ships, a distinction between assembly and embarkation (to boats or rafts) is made. These are separate from each other. The decision whether to enter the boats or rafts is usually made after assembly is completed [1].

8.4.6.3 Life-Saving Equipment in EER

1. **Life boats:** These are used for emergency evacuation from the platform fire retardant, self-righting lifeboats are provided on all installations. Their location is supposed to be known to each individual on board the installation. In case of any emergency, the instruction of boat supervisor/captain must be followed without any delay/hesitation (Figure 8.8).
2. **Life raft:** Inflatable life rafts are provided on platforms as a backup for lifeboats. These are also placed at well platforms where there is no lifeboat. Life rafts are positioned strategically and are easy to operate. Life rafts are inflated automatically when thrown into the sea by an automatic mechanism.
3. **Life buoys:** These are round floatation rings fitted with an automatic light. These rings are used for saving life when somebody accidentally goes overboard.

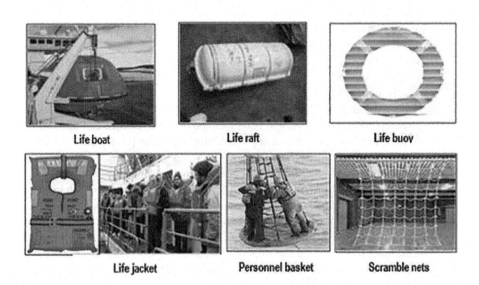

FIGURE 8.8 EER equipment [1].

Fire blanket　　　　　　Breathing air apparatus　　　　　Fire suits

FIGURE 8.9 EER equipment [1].

4. **Life jackets:** These are floatation jackets worn for floatation.
5. **Personnel baskets:** These are specially designed and certified baskets for personnel transfer from boat to platform or boat to boat using crane.
6. **Scramble nets:** These are specially designed nets placed strategically on various decks and used for climbing down in an emergency.
7. **Fire blankets:** These are used for escaping a blaze.
8. **Breathing air apparatus:** Breathing apparatus is used to escape smoke or H_2S emergencies (Figure 8.9).
9. **Fire suits:** These are specially designed suits used for firefighting. The operating instructions and the maintenance procedures are always attached with these items. All personnel on board must be familiar with such equipment, and location should be easily accessible and prominent.
10. **Thermal protective aid:** It is a bag or suit made of waterproof material with low thermal conductance.
11. **Immersion suit:** It is a protective suit which reduces body heat loss of a person wearing it in cold water.
12. **Totally enclosed lifeboats:**
    a. *Float-free launching:* Launching a survival craft whereby the craft is automatically released from sinking ship.
    b. *Free fall launching:* Launching a survival craft whereby the craft with its complement of persons and equipment on board is released and allowed to fall into the sea without any restraining apparatus [1] (Figure 8.10).

- Lifeboats with self-contained air support systems must have a supply of safe and breathable air, provided at slightly above atmospheric pressure to keep out fumes, for a shutdown period of not <10 minutes. Air bottles are placed beneath the seats at the bottom of the boat to meet this requirement.
- Fire-protected lifeboats must protect their occupants from the effects of a continuous oil fire that envelopes the boat for a period not <8 minutes. Protection for the hull is normally provided by water spray systems which is supplied with water by a self-priming, motor-driven pump connected to the main engine. An "on-off" valve is provided on the pump's suction pipe to ensure that the water spray can be turned off when not required.

Offshore Health, Safety, and Environment   225

FIGURE 8.10 Free-fall launching [1].

- Propulsion: Completely enclosed lifeboats must be capable of a speed of at least 6 knots and have sufficient fuel to maintain this speed for not <24 hours. They must also be able to tow a loaded 25 people life raft at a speed of at least 2 knots, if required [1].

8.5 NAVIGATION AIDS

A navigational aid is a marker that aids the traveler in navigation, usually nautical or aviation travel. Common navigation aids include lighthouses, buoys, fog signals, and day beacons. Figure 8.11 displays the different types of navigational aids utilized during offshore operation and engineering.

To avoid potential hazards caused by to and fro movements of ship navigation aids are utilized. Navigational aids are used to:

a. avoid dangerous zones like offshore production risers/gas pipelines
b. locate ports during night and bad weather condition
c. follow proper vessel approaches

The different types of navigational aids include:

1. Fixed-type
   a. Lighthouse: It is a tower structure built of masonry or reinforced concrete. Beacon light is provided. Tower is divided into a number of floors and is made strong enough to withstand heavy wave motion. Usually, it is designed in such a way that the lights are visible up till 30 km.

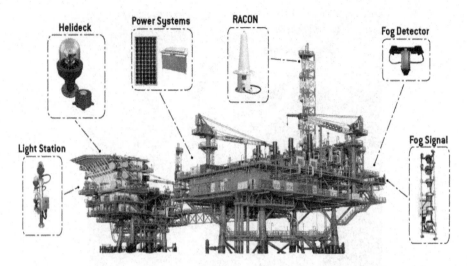

FIGURE 8.11 Different navigational aids on an offshore structure (Tideland Offshore Systems) [7].

    b. Beacon lights: Beacon light is fixed or flashing for easy identification by vessels. It is used in groups for alignment or change in direction.
2. Floating type:
    a. Buoys: These are small-sized floating structures, generally in the form of large cylindrical cans and drums.
        – Different types of buoys are:
          1. Luminous boys
          2. Audible buoys or bell buoys
          3. Mooring buoys
    b. Lightships are used in locations where it is not practical to build lighthouses, small ships are used for the purpose. The bulls of lightships are generally painted with red color. They are generally held in position by anchoring or mooring.

8.5.1 Emergency Position-Indicating Radio Beacon (EPIRB)/ Search and Rescue Transponder (SART)

- These are present on all ships, other than those operating within vhf range of coast stations and equipped with vhf EPIRB (currently operating on 121.5 MHz, a float-free satellite EPIRB.) This operates in the 406-MHz band, which allows for its location by polar orbiting satellites of the COSPAS-SARSAT systems. The position of beacon can be determined by analyzing the Doppler shift of radiofrequency which occurs when satellite passes within its transmitting range [8].

Offshore Health, Safety, and Environment

- The EPIRB signal includes a short-coded message which provides SAR authorities information concerning the type of beacon, the manufacturer's serial, and whether it has any extra homing devices fitted to it, such as a radar transponder (9 GHz) or VHF transmitter (121.5 MHz or VHF 16) [8].
- Information on 406 MHz can be stored and processed aboard the satellite and then transmitted down to the next receiving ground station. Although in some areas this may mean a delay of 2 or 3 hours before the rescue authorities are made aware of a beacon alert, it does give the system virtual worldwide coverage. An EPIRB must be capable of being manually operated or of operating automatically after breaking free of the vessel after immersion to a depth of not more than 4 m [8].
- It is capable of transmitting continuously for a period of not <48 hours and has batteries that do not need replacement at more than yearly intervals. It is of a highly visible color, fitted with reflecting material, and is capable of floating upright in calm water. An indicating light is incorporated in it to show the user that the EPIRB is transmitting signals [8].
- Recent advancement in navigation aids include [7]:
  1. Lighted beacons and floating aids.
  2. Radio aids and automatic identification system (AIS)
  3. Audible aids and power sources
  4. Support structures

In addition, pyrotechnics with the help of parachutes are utilized for navigation and safety purposes.

8.5.2 PYROTECHNICS

- All pyrotechnics have a storage life of 3 years and display their date of manufacture and expiry date. Out-of-date pyrotechnics should never be used for testing or practice purposes, as there is a possibility that the chemicals may have deteriorated with age. If expired, pyrotechnics have to be disposed off at sea; they must be weighted and dumped in deepwater well away from land. If this is not done, there is always a possibility that they will be washed away ashore and cause injury to the members of the public, especially children, who may tamper with them [8].
- At least 12 should be present on passenger and cargo ships, in addition to those required aboard survival craft. Four are carried in most lifeboats and two are carried in life rafts. Single red flare suspended on a parachute. Luminous intensity not <30,000 candelas. Burn time at least 40 seconds. Altitude attained at least 300 m. Descend rate not more than 5 m/s. Approximate range 20–25 miles [1] (Figure 8.12).

FIGURE 8.12 Rocket parachute flares used in navigation aids [8].

8.6 WELL INTEGRITY

Well integrity is one of the factors that play a major role in preventing any kind of hazard on offshore or even on onshore platforms. Well integrity depends on the following features, which include well barriers and operational decisions during an event:

- Design flaws when long-term effects are not sufficiently considered.
- Improper selection of tubing, casing, or any other well equipment during drilling and well completions.
- Inadequate use of most updated techniques during well completions.
- Operational decisions during abnormal situations often lead to well failure.

The challenge is to account for rare events that may lead to major incidents. The normal approach is to focus on frequent and low-consequence incidents.

Well barriers are equipment that are utilized during drilling and well completions. These prevent the uncontrolled flow of HC's into the well. The clear-cut definition and description of each well barrier is made before commencement of an activity or operation as per the specific acceptance criteria of well barrier elements (WBE). Well barrier acceptance criteria are technical, and operational requirements need to be fulfilled to certify the well barrier or WBE for its intended use [9].

- Most integrity problems are within barrier elements such as tubing, annulus safety valves (ASVs), casing, cement, and wellhead. Some of the well integrity problems faced and the category of barrier element failure are listed below [9].
  - Some tubing problems are leakage in production tubing above the downhole safety valve (DHSV), tubing-to-annulus leakage, or internal leak in tubing-hanger neck seal.
  - Some of the problems with ASVs are ASV leakage or ASV failure.
  - Casing problems such as casing leakage (connections that are not gas-tight) or collapsed casing.

Offshore Health, Safety, and Environment

- Cement problems such as no cement behind the casing and above the production packer, leaks likely along cement bonds, or leak through cement microannulus.
- Wellhead problems include leakage in wellhead from Annulus A to Annulus B because of a wrong seal type in the wellhead [9].
- Tubing leakage is a dominant factor in failure (39%) in wells aged 0 to 19 years. Wells aged 0 to 14 years have barrier element issues such as tubing, ASV, and cement. Wells aged 15 to 29 years have barrier element issues such as tubing, casing, and pack-offs [9].

Thus, well integrity deteriorates as well barrier deteriorates with time, and therefore, caution should be taken during designing and installation of each well barrier, otherwise the following well failures can occur.

8.6.1 Well Failure: Example 1

During workover of a well on a production platform, the load-bearing surface casing collapsed, resulting in a wellhead that dropped onto the platform structure. The root cause was severe corrosion near the top of the surface-casing annulus. Corrosive seawater gained access through a cement port that was left open during the drilling phase. Temperature and tidal variations through platform shaft leaks were accompanying factors. The particular well was shut for approximately 1 year before production commenced [9].

8.6.2 Well Failure: Example 2

The operator had installed a specially designed "slim" wellhead in a field. The casing hanger had only 8° taper as opposed to the usual 40°. During pressure testing, the casing hanger was pulled through the wellhead. Later, the operating company experienced a similar incident with a tubing hanger. The root cause of the failures was axial overload because the slim wellhead tapers had limited capacity. The wellhead capacity had been uprated from 350 tons to 600 tons. The investigation report, however, showed that the manufacturer test had failed, leaving it incorrect to allow such an uprating. After these incidents, the operating company limited the load to the initial design value. Another finding in this case was that the operator did not have in place an updated well design manual. This was corrected after the audit [9].

Figure 8.13 explains the severity of incident which occurred due to well integrity failure in 2010. This is one of the biggest incident which resulted in a huge loss of lives and money, as well as environmental damage. This also raised the question about the seriousness of companies during offshore operations.

Thus, we need to ensure best practices and update techniques of design, installation, and operational decisions during offshore operations. The failure to do so can result in costliest and deadliest incidents. The next section discusses the costliest and deadliest incidents which have occurred in the world during offshore operations and engineering.

FIGURE 8.13 Well integrity problems resulting in the Macondo blowout (Deepwater Horizon Incident) [10].

8.7 LIFE EXTENSION AND ASSESSMENT OF OFFSHORE STRUCTURES

It is essential to conduct assessment of offshore structures to avoid untoward incidents like failure of member of structure or structural collapse or any such incident in offshore.

8.7.1 STRUCTURAL COLLAPSE

Structural collapse can also pose a major hazard, and thus, extreme caution is taken during its design and selection of the materials for the structure. Extreme care is taken during installation to avoid any mishandling as it may have a bigger consequential effect. To ensure strength of the structures, simulation of 100 years is conducted under assumptions of extreme conditions of offshore environment. If proper caution is not taken, then structures can collapse, as shown in Figure 8.14. It is essential to assess integrity of structure to avoid any such incident.

An increasing number of offshore platforms are reaching their design life. For various reasons, these platforms will require an assessment of their structural integrity. For example, one may experience operational changes of the platform that may lead to increased loads or there may be damages that reduce the structural capacity. Consequently, the design premises may have changed significantly and may result in increased uncertainty about the safety of the structure. In such cases, the assessment process focuses on reassuring that the structure has adequate safety [4].

Design standards are based on theories, methods, and experience for structures in a given design life (e.g. fatigue design and corrosion protection design). Sound methods are required for better understanding the life and strength of structures. Such methods can normally be "operating condition-based design", where inspection, maintenance, and repairs are included in the assessment in an integrated manner [4].

Figure 8.15 illustrates the assessment process during life execution of the structures.

Offshore Health, Safety, and Environment   231

FIGURE 8.14 Offshore structure collapse [11].

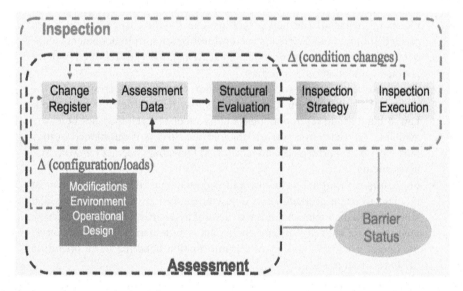

FIGURE 8.15 Assessment for life execution of offshore structures [4].

8.8 ASSESSMENT PROCESS

The process for assessment of an existing offshore structure is described in ISO 19902 (2007), and may be illustrated as shown in Figure 8.15. It illustrates how inspection and assessment play a major role in checking the status of various safety barriers [4]. The following challenges are encountered in reliable assessment:

- Quality of data
- Proof of structural integrity with increased loads
- Capacity and performance of damaged structures
- Extended life

8.8.1 Quality of Data

- Lack of documents, lack of reliable updation of data during the lifecycle of operations and structures, lack of information, lack of awareness, and sometimes data manipulation results in unreliable and incorrect assessment.
- Compared with ordinary design work, the task of assessing an existing structure is often followed by uncertainties about important structural parameters. Even if design documents were produced during the construction phase, often there are cases where not all of the required documents are available and their questions about their correctness. As built documents are not always produced, there are cases where important information is missing in the documents.
- Most structures used by the oil industry undergo small or large modifications during the service life of the installation, and therefore, it is often necessary to consult a change register to check what documents comprise relevant information about the current structure. Depending upon the complexity of the installation and the completeness of the change register, it can prove difficult to read correct structural information from documents without performing on-site inspections. Measurements may be required both for geometry and material properties.
- The condition of the structure is another source of uncertainty for the structural assessment. Structural surveys may be insufficient for the problem at hand or survey reports, especially older reports, may not be easily available. In some cases, the accuracy of the surveys is questioned as they may have been performed with a different focus than needed for structural assessments.
- For complex installations having a long operational life, it will prove most efficient to use a product-based data storage system for swift access and storage of all data relevant for the structural assessment. An example is sesame structure integrity system where data is systematically stored to assist the operator with all aspects of the information handling when operating offshore structures [4].

8.8.2 Proof of Structural Integrity with Increased Loads

- The consequences of a structure failing a check for structural integrity at the design stage will often mean a change in the dimensions on the drawing and marginal increase in material and fabrication cost. Even during the construction phase, there are possibilities to perform reasonable large reinforcements without considerable cost or schedule impacts. During operation, even small structural modifications may be of significant cost, especially if it involves production shutdown or underwater work.
- Therefore, there are strong incitements to remove unnecessary conservatism in checking the structural integrity of existing structures. Detailed knowledge of all topics is relevant for the assessment of offshore structures from consideration of the environmental parameters and their combinations,

Offshore Health, Safety, and Environment 233

load calculations, and the determination of structural resistance represent a resource to avoid costly reinforcements or premature shutdown of the installation. This means that traditionally adopted design methods, as presented in design codes, need to be supplemented by more refined methods to achieve optimum operations of the structure [4]. Thus, periodic updation of the design methods with technology advancement is required for a reliable assessment process.

8.8.3 CAPACITY AND PERFORMANCE OF DAMAGED STRUCTURES

- Structures that are damaged due to ship collisions, dropped objects, extreme temperatures from fires, or from degrading processes like corrosion and fatigue will loose their structural load-carrying capacity. Similarly, as for the situation with increased loads, it is always beneficial to check if the requirements regarding structural integrity are met without performing repair work. Thus, we need to be more careful and meticulous during selection of methods for the best assessment of damaged structures.

8.8.4 EXTENDED LIFE

- The time-dependent degrading processes such as corrosion and fatigue may result in threat against the structural integrity of a structure. Structural integrity may need to be reviewed for extended life due to modification/ changes carried out during the lifetime on account of change in process parameters and/or environmental conditions. Consequently, the structure has to be designed to maintain its integrity for a certain time called the design life. Within the design life, the degrading processes is assumed not to impair the structural integrity provided that the design presumptions are fulfilled. The presumption includes maintenance of corrosion protection systems and inspections to detect possible fatigue cracks. The structural integrity of the structure needs to be revisited to extend the designed life of offshore process facilities for continued operation.
- In the simplest case of assessment, one may find that the corrosion protection system is adequately protecting the structure, and that any material loss or additional loads put up due to modification is within the originally defined limits, and that a new calculated fatigue life is sufficient for an extended period. However, in other cases, it may be necessary to consider larger material loss than planned due to corrosion, and that the calculated fatigue life is less than required. In these cases, it will be possible to prove adequate structural integrity by using more refined methods for assessing the structural capacity. This is valid both for the presence of corrosion outside the original limits and for structures with cracks. More refined methods can also change the conclusions about the remaining fatigue life.
- However, even without proving a theoretical fatigue life, safe operations of the structure can be achieved by the knowledge gained from the in-service experience of the structure. The information of the behavior of a platform

234 Offshore Operations and Engineering

that for a long time has been regularly inspected is of large value to prove that a structure may have a practical fatigue life that surpasses the theoretical fatigue life. By conducting adequate and carefully planned inspection, a structure may be safely operated considerably longer than what is predicted by theoretical fatigue analyses [4].

8.9 IMO RESOLUTIONS

- The IMO has adopted a number of resolutions in relation to the safety and protection of offshore oil installations, particularly with respect to safety zones around such installations. Resolution A.341 (IX) contains recommendations on the dissemination of information, charting, and manning of drilling rigs, production platforms, and other similar structures [12]. Resolution A.379(X) provides a recommendation for the establishment of safety zones in offshore exploration areas. Resolutions A.621 (15) and A.671 (16) contain certain provisions in relation to safety zones and to prevent the infringement of safety zones. These two resolutions will be discussed below [13].

8.9.1 RESOLUTION A. 621(15)

- The Resolution recommends that vessels which are passing close to offshore installations or structures navigate with care when passing near offshore installations, take early and substantial avoidance action when approaching such installations, use any designated routing systems established in the area, and maintain a continuous listening watch on the navigation bridge on VHF channel 16 when navigating near-offshore installations to allow radio contact to be established between installations, vessel traffic services, and vessels [12].
- The Resolution further recommends that the coastal state which has authority and jurisdiction to regulate the use and operation of offshore installations issue early notices to mariners by appropriate means to advise vessels of the location or intended location of offshore installations or structures, the breadth of any safety zones, and the rules which apply therein. It is also recommended that the coastal state require operators of offshore installations to take adequate measures, such as effective lights and sound signals, to prevent the infringement of safety zones around such offshore installations or structures. Finally, it is recommended that the coastal state request operators of offshore installations to report actions by vessels which jeopardize safety, including the infringement of safety zones [13].

8.9.2 RESOLUTION A. 671(16)

- The Resolution recommends that governments consider where traffic patterns warrant the establishment of safety zones around offshore installations or structures [14]. It is recommended that governments take all the necessary steps to ensure that, unless specifically authorized, ships flying their flags do not enter or pass through duly established safety zones. The Annex of the Resolution further provides certain recommendations with respect to the dissemination of information related to offshore installations and structures. For example, the coastal state should be responsible for the dissemination of information essential for the safety of navigation or any other legitimate activities within the area in which, in accordance with international law, it has sovereign rights and jurisdiction [12]. Details of any safety zone around the installation or structure and any fairways and routing systems established in its vicinity including, where relevant, their marking [12], should be taken into account to deal with the dissemination of information. Any features of a sufficiently permanent nature, such as permanent installations or structures, bottom obstructions, pipelines, navigational mark sand prohibited areas, should be shown on all appropriate navigational charts [13].

8.10 OFFSHORE FACILITIES INTERFERENCE

8.10.1 PROTECTION OF OFFSHORE FACILITIES/RIGS

- Offshore oil rigs/facilities may collide with ships and other sea-going structures. This may lead to loss of life, sea pollution, and economic damage. For example, during 1973 to July 1995, a total of 463 incidents of vessels colliding with offshore oil installations were recorded on the UK continental shelf. Offshore oil installations may also be damaged by the acts of other users of the sea [15]. They further may be subject to military and/or terrorist attack. The International Maritime Organization (IMO) has adopted certain regulations to ensure the safety of oil rigs/facilities and to prevent the infringement of safety zones around offshore installations or structures. The establishment of a safety zone around oil rigs/facilities is one of most effective ways to protect them from collision and/or other dangers.
- A limit of 500 m for safety zones around oil rigs/facilities was concluded at UNCLOS III because different states could not agree on anything else. Although, according to the LOSC, a wider safety zone, as authorized by generally accepted international standards or as recommended by the competent international organization, can be established, no recommendation on the extenuation of safety zones beyond 500 m has been made by the competent IMO. Safety zones may be established around both mobile and fixed oil platforms.
- It seems that the establishment of safety zones around oil rigs/facilities is not a sufficient measure to prevent collision between oil installations and ships. The statistics for collisions between oil rigs and ships in the North Sea indicate that the 500 m safety zones have not been effective enough to prevent collisions [13].

8.11 COSTLIEST AND DEADLIEST EVENTS IN OIL AND GAS INDUSTRY

8.11.1 Bohai 2 Oil Rig Disaster, China (1979)

- The Bohai 2 oil rig disaster in the Gulf of Bohai off the coast of China in November 1979 is the sixth most fatal offshore oil rig disaster. It caused the death of 72 out of 76 people on board as the Bohai 2 jackup rig capsized and toppled in sea.
- The accident was caused by a storm which occurred while the rig was being towed. Fierce winds broke the ventilator pump of the platform causing a puncture hole in the deck resulting in extensive flooding.
- The loss of stability due to flooding coupled with severe weather conditions eventually led to the capsizing of the jackup. The accompanying tow boat could not rescue the crew members, who were also believed to lack proper training on emergency evacuation procedures and the use of lifesaving equipment.
- The post-disaster investigations reported a failure in correctly stowing the deck equipment prior to towing. It was also reported that standard tow procedures were not followed given the bad weather conditions. The sunken jackup rig was eventually salvaged with explosives by the Yantai Salvage Company in April 1981 [13,15].

8.11.2 Alexander L. Kielland, North Sea, Norway (1980)

- The Alexander L. Kielland was a semi-submersible platform accommodating the workers of the bridge-linked Edda oil rig in the Ekofisk field, about 235 miles east of Dundee, Scotland, in the Norwegian continental shelf. The platform, operated by Phillips Petroleum, capsized in March 1980, killing 123 people.
- Only 89 out of the 212 workers survived the accident and most died by drowning as the platform turned upside down in deep waters. The platform capsized after the failure of one of the bracings attached to one leg of the five-legged platform structure after strong winds created waves of up to 12 m high on the day of the accident.
- Once the first broke, all bracings attached to the leg failed in succession, the platform lost one of its five legs, and the entire structure tilted 30°. Five of the six anchor cables snapped but the platform was stabilized for some time by the remaining single cable, which ultimately snapped.
- Official investigation concluded that the root cause of the accident was an undetected fatigue crack in the weld of an instrument connection on the bracing [13].

8.11.3 Ocean Ranger Oil Rig Disaster, Canada (1982)

- The Ocean Ranger oil drilling rig disaster which occurred in the North Atlantic Sea off the coast of Newfoundland, Canada, on 15 February 1982 is one of the deadliest offshore oil rig accidents in history. The offshore oil drilling capsized and sank, killing 84 crew members on board.

Offshore Health, Safety, and Environment 237

- The semi-submersible mobile offshore drilling rig owned by Ocean Drilling and Exploration Company (ODECO) was hired by Mobil Oil, Canada for drilling exploration well at Hibernia field at the time of the accident. One of the biggest rigs built by then, the 25,000 ft rig was 396 ft long, 262 ft wide, and 337 ft tall, with the capacity to operate 1,500 ft beneath water and drill up to a depth of 25,000 ft below the seabed.
- The rig capsized due to a very strong storm which produced 190 km/h winds and up to 65 ft (20 m) high waves. On February 14, 1982, it was reported that a port hole window had broken and water had entered into the ballast control room of the Ocean Ranger. The ballast control panel was noticed to be malfunctioning 2 hours later.
- Crew workers abandoned the rig and moved into lifeboat stations but only one lifeboat with 36 crew members inside could be launched successfully. At least 20 crew were reported to be in the water before the rig sank between 3:07 am and 3:13 am after staying afloat for about one and half hours [13].

8.11.4 Glomar Java Sea Drillship Disaster, South China Sea (1983)

- The Glomar Java Sea Drillship disaster which occurred on 25 October 1983 in the South China Sea caused the death of 81 people when the drillship capsized and sank at a depth of 317 ft about 63 nautical miles south-west of Hainan Island, China, 80 nautical miles east of Vietnam.
- The 5,930 ft Glomar Java Sea drillship was built by the Levingston Shipbuilding Company of Orange, Texas, in 1975 and delivered to Global Marine. The 400 ft long drillship was contracted to ARCO China at the time of the disaster. The vessel had performed drilling for ARCO in the Gulf of Mexico between 1975 and 1881 and operated off the coast of California for some time before its arrival in the South China Sea in January 1983. Operations ceased prior to the arrival of tropical Storm Lex as it approached from the east of the drilling site. Global Marine's office in Houston, Texas reported that the drillship was experiencing 75 kt (138.9 km/h) winds over the bow, but the contact was abruptly lost.
- No survivors were found in the extensive search operation conducted thereafter. The wrecked drillship was found in an inverted position 1,600 ft southwest of the drilling site. Only 36 bodies were found and the remaining 45 crew members were presumed dead [13].

8.11.5 Enchova Central Platform Disaster, Brazil (1984)

- The Enchova Central Platform disaster in the Campos Basin near Rio de Janeiro, Brazil, killed 42 people in August 1984. The accident occurred due to a blowout which caused a fire and explosion at the central platform of the Enchova field operated by Petrobras.
- Most workers were evacuated from the platform by lifeboats and helicopter except for 42 workers who lost their lives during the evacuation process. Malfunctioning of the lowering mechanism of a lifeboat caused

238 Offshore Operations and Engineering

the death of 36, while 6 died as they jumped from the platform into the sea. The lifeboat remained vertically suspended because of the failure of the bow hook and eventually fell 20 m deep into the sea as its supporting cables snapped.

- Another disaster struck the Enchova platform on April 24, 1988 as one of its 21 wells blew out and eventually ignited. The well suffered a blowout while undergoing a workover to convert it from oil production to gas production. The fire caused by the blowout on the platform led to massive damage topside, however, all the workers were safely evacuated to the nearby floating accommodation ship without a single casualty.
- The platform remained on fire for a month and Petrobras eventually drilled two relief wells to control the blowout. The platform was declared a total loss and replaced by a new facility that commenced production nearly 18 months after the accident [13].

8.11.6 Piper Alpha, North Sea, United Kingdom (1988)

- The Piper Alpha disaster in the North Sea, UK, which killed 167 people in July 1988, is the deadliest offshore oil rig accident in history. Discovered in 1973 and brought on-stream in 1976, Piper Alpha was one of the biggest offshore oil platforms in the UK producing more than 300,000 barrels of crude a day (about 10% of the country's total crude production). The offshore platform started producing gas in the early 1980s and had three main gas transport risers and an oil export riser before disaster struck, destroying the entire facility causing an estimated loss of $1.4 bn.
- The Piper Alpha disaster occurred due to gas leakage from one of the condensate pipes at the platform on July 6, 1988. The pressure safety valve of the corresponding condensate-injection pump was removed during the day as part of the routine pump maintenance. The open condensate pipe was temporarily sealed with two blind flanges. The temporary disc cover, however, remained in place during shift change in the evening as maintenance work was not complete. The condensate-injection pump was not supposed to be switched on under any circumstances.
- Communication errors, however, led the night crew staff at the platform to turn on the pump after the other pump tripped. It resulted in leakage of gas condensate from the two blind flanges causing gas ignition and serial explosions on the platform. Only 61 out of the 226 workers survived the disaster and it took close to 3 weeks to control the fire.
- At the time of the disaster, the platform was managed by Occidental in block 15 of the UK Continental Shelf about 120 miles northeast of Aberdeen [13].
- A committee was formed to understand the reasons of the incidents and following were their findings [16]:
  - Failure of permit to work system.
  - No formal handover from day shift to night shift.
  - Non-compliance to company procedures.

Offshore Health, Safety, and Environment 239

- Company management was easily satisfied with the safety system (lack of control).
- No proper training; safety policies were in place but not practiced.
- Emergency induction was not provided or inconsistently given.
- No drills or exercises were conducted to test emergency preparedness.
- No emergency response training was provided.
- Failure to conduct risk assessment.
- Inadequate guidance or means to assess the effectiveness of the safety management system.
- Poor management system [16].

8.11.7 SEACREST DRILLSHIP DISASTER, SOUTH CHINA SEA, THAILAND (1989)

- The Seacrest Drillship disaster in the South China Sea 430 km south of Bangkok, Thailand killed 91 crew men on the November 3, 1989. The 4,400 ft drillship was anchored for drilling at the Platong gas field, owned and operated by Unocal. The drillship was capsized by the Typhoon Gay which produced 40 ft high waves on the day of the accident.
- The Seacrest drillship, also known as The Scan Queen, had been operational in the Gulf of Thailand since 1981 as a drillship for Unocal. The drillship was reported missing on November 4, 1989, and was found floating upside-down by a search helicopter the next day. The capsize was believed to be so quick that there was no distress signal and no time for the crew members to respond to the disaster.
- Just 6 out of the 97 crew members on board were rescued by fishing boats and the Thai Navy. Apart from the severe weather conditions, the seaworthiness of the drillship was questioned as the likely cause for the tragedy.
- The ship also had 12,500 ft of drillpipe in its derrick, which was believed to have resulted in an unstable high center of gravity [13].

8.11.8 MUMBAI HIGH NORTH DISASTER, INDIAN OCEAN (2005)

- The Mumbai High North disaster on 27 July 2005 in the Arabian Sea, around 160 km west of the Mumbai coast, killed 22 people. Mumbai High North, one of the producing platforms of the Mumbai High field owned and operated by India-owned Oil and Natural Gas Corporation (ONGC) caught fire after a collision with the multipurpose support vessel (MSV), Samudra Suraksha.
- Strong swells pushed the MSV toward the platform hitting the rear part of the vessel causing rupture of one or more of the platform's gas export risers. The resultant gas leakage led to ignition that set the platform on fire. Heat radiation also caused damage to the MSV and the Noble Charlie Yester jackup rig engaged in drilling operation near the platform.
- The accident caused significant oil spill and a production loss of about 120,000 barrels of oil and about 4.4 million cubic meters of gas a day. ONGC opened a new platform at Mumbai High North in October 2012 [13].

8.11.9 Usumacinta Jackup Disaster, Gulf of Mexico (2007)

- The Usumacinta Jackup disaster on October 23, 2007 in the Gulf of Mexico claimed 22 lives after a collision with the PEMEX-operated Kab-101 platform in the Bay of Campeche.
- The Usumacinta Jackup was positioned alongside the Kab-101 platform to complete drilling of the Kab-103 well. A storm with winds of 130 km/h and up to 8 m of waves created an oscillating movement, which eventually caused its cantilever deck to hit the production valve tree on the Kab-101 platform.
- The collision resulted in oil and gas leakage leading to the closure of the safety valves of two production wells at the platform. The crew members were, however, unable to seal the valves completely, which resulted in continued leakage of oil and gas, which eventually ignited causing fires on the platform. Twenty-one people were declared to have died during the evacuation, and one worker missing in the rescue operation was presumed dead.
- The Usumacinta Jackup also suffered two more fire breakouts in the following month during well control operations. The fire was, however, extinguished without any loss of life, and complete control of the well was achieved by the middle of December 2007. Approximately 5,000 barrels of oil were reported to have lost from the well without being recovered [13].

8.11.10 Deepwater Horizon, Gulf of Mexico (2010)

- On April 20, 2010 the Mobile Offshore Drilling Unit, The Deepwater Horizon was completing drilling operations at the Macondo Well located 50 miles off the coast of Louisiana in the Gulf of Mexico.
- During the drilling operations aiming to temporarily abandon the well (for subsequent completion as a subsea producer), there was a loss in well control resulting in an influx of formation fluids known as a kick, which eventually resulted in the release of voluminous amounts of liquid and gaseous hydrocarbons.
- The culmination of the hazardous volume of the same resulted in explosion claiming the loss of 11 lives. The rig eventually sank and leading to the loss of the Deepwater Horizon and resulting in continuous oil spill for the next 87 days. This incident is popularly known now as the British Oil Spill owing to a huge environmental catastrophe in the history of oil and gas.
- The event occurred due to lack of best practices followed by BP during well completions, and finally BP was heavily penalized for this negligibility by the court of law in United States.

8.12 OFFSHORE SECURITY THREATS

Attacking offshore oil and gas installations is not a new phenomenon. In fact, the first attack on an offshore oil installation took place more than 100 years ago on August 2, 1899 off the shores of Santa Barbara, California. In the last 25 years

Offshore Health, Safety, and Environment

there have been about 50 attacks and security incidents involving offshore installations. These were carried out by various perpetrators with different motivations, objectives, tactics, and capabilities. These include terrorists, insurgents, pirates, criminal syndicates, environmental activists, anti-oil activists and other types of protesters, hostile nation-states, and sometimes other unknown groups and individuals [17]. Any unlawful interference with offshore oil and gas operations or an act of violence directed toward offshore installations is considered an "offshore security threat:. Offshore security threats may be classified in several ways based on different criteria. One such classification is based on geographical criteria, such as local or global, national or transnational. The attacks may come from various sources: individuals or groups, internal or external to a state, or a combination of both. This article categorizes threats faced by offshore oil and gas installations based on the type of activity [18].

8.12.1 PIRACY

- Piracy is one of the most visible security threats to offshore oil and gas installations, as exemplified by activities in the Gulf of Guinea. In the last 7 years at least six pirate attacks have been reported worldwide. Four of these took place in the Gulf of Guinea (the April 1, 2007 attack on Bulford Dolphin mobile offshore drilling rig; May 3, 2007 attack on FPSO (floating production, storage, and offloading vessel) Mystras; May 5, 2007 attack on Trident VIII offshore rig; January 5, 2010 attack on FSO Westaf). One attack took place near India in 2007 (the March 22, 2007 attack on Aban VII jackup rig) and one near Tanzania in 2011 (the October 3, 2011 attack on Ocean Rig Poseidon drill ship). Apparently, there was another attack off Tanzania on a drillship contracted to Ophir Energy in September 2010; however, that attack is not reported in the IMO reports on piracy [18].

8.12.2 TERRORISM

- Terrorism is another security threat to offshore oil and gas installations. To date, there have been only two terrorist attacks against offshore installations. On April 24, 2004, in Iraq, Iraq's Al Basrah Oil Terminal (ABOT) and the Khawr Al Amaya Oil Terminal (KAAOT) in the Persian Gulf were attacked nearly simultaneously by suicide boats; these attacks were allegedly carried out by the Al-Qaeda-affiliated Zarqawi network based in Iraq. Although largely unsuccessful from a physical asset standpoint, they did result in three fatalities and closure of the terminals for about a day, consequently resulting in loss of revenue due to a production shutdown [18].

8.12.3 INSURGENCY

- Insurgency is motivated by political struggle and insurgents often causing destruction, damage, and casualties to offshore installations. Insurgency groups are responsible for about one-third of attacks and security incidents involving offshore installations, most of which occurred in the Gulf of Guinea. For example, between 2006 and 2010, the Movement for the Emancipation of Niger Delta (MEND) insurgency group carried out at least 13 attacks on offshore oil and gas installations in the Niger Delta region of Nigeria as part of their campaign against the oil and gas industry to achieve fair distribution of oil profits and compensation from oil companies. These include the attack on the Bonga FPSO by militants about 90 nautical miles offshore on June 19, 2008 and the bombing of Forcados offshore oil loading terminal on June 29, 2009 [18].

8.12.4 ORGANIZED CRIME

- Organized crime can also interfere with offshore oil and gas operations. The types of crimes relevant to the oil and gas industry include oil theft, extortion, armed robbery, theft of property, and other forms of criminal profiteering. There have been at least two reported attacks on offshore installations involving organized criminal groups. These are the attack on the mobile offshore drilling rig Allied Centurion in Malaysia on December 26, 2008, where a group of armed robbers boarded the drilling rig and stole stores of goods and property. Another incident that can be attributed to organized crime is the attack on the offshore Moudi oil terminal in Cameroon on November 17, 2010 by a "hybrid criminal/separatist" group called Africa Marine Commando (AMC), apparently for failure to pay a "security tax" previously demanded by the perpetrators [18].

8.12.5 CIVIL PROTEST

- Civil protest also poses a security threat to offshore oil and gas installations. Interferences with offshore operations can be caused by non-violent environmental activists, indigenous activists, labor activists, striking workers, and anti-government protesters. Greenpeace activists have caused interferences with operations of offshore installations on several occasions, including an attempt to board an oil rig about 170 nautical miles off the coast of Massachusetts in the United States on July 25, 1981, unauthorized boarding and occupation of Shell's Brent Spar floating offshore oil storage facility in the North Sea on April 30, 1995, an unauthorized boarding of the Stena Don offshore drilling rig off the coast of Greenland on August 31, 2010, interference with operations of the Stena Carron drillship in the waters off the Shetland Islands northeast of mainland Britain in the UK on September 21, 2010, and the unauthorized boarding of LeivEiriksson offshore drilling rig in Turkish waters. There have been other offshore security incidents caused by civil protest such as an unauthorized boarding of Parabe offshore oil production platform by protesters in Nigeria on May 25, 1998, and the seizure of offshore installations by striking oil workers off the coast of Nigeria in April 2003 [18].

Offshore Health, Safety, and Environment

8.12.6 Interstate Hostilities

- Certain conduct or actions of nation-states may also represent a security threat to offshore installations. These can be in the form of interstate armed conflicts and wars, maritime boundary disputes, and state terrorism. There have been at least six security incidents involving actions of hostile states including the attack in March 1983 by Iraqi planes on the Iranian offshore platform at the Nowruz oil field; the October 19, 1987 attack on the Iranian R-7 and R-4 offshore oil platforms in Reshadat and the April 18, 1988 attack on Iranian offshore oil complexes, Salman (aka Sassan) and Nasr (aka Sirri) by US military. Other examples are the June 3, 2000 incident where the Suriname Navy ordered American-owned and operated offshore oil drilling rig, CE Thornton, to stop drilling and leave the area and threatened to use force if its demands were not complied [18].

8.12.7 Vandalism

- Vandalism is another security threat to offshore oil and gas installations. Vandalism can be referred to as "damaging cargo, support equipment, infrastructure, systems or facilities" (Australian Government, Department of Transport and Regional Services, Offshore Security Assessment Guidance Paper (2005): 15). This category of threat can include violent actions of radical environmental and animal rights groups and acts by members of local populations intended to cause damage to company property. For example, on August 2, 1899, in the United States, when an oil company began to construct an oil derrick off the shores of Montecito, an affluent suburb of Santa Barbara, California, a local mob attacked the rig and demolished it (Harvey Molotch, William Freudenberg and Krista Paulsen, "History Repeats Itself, But How? City Character, Urban Tradition, and the Accomplishment of Place" American Sociological Review 65 (2000): 804) [18].

8.12.8 Internal Sabotage

- Internal sabotage is also a potential security threat to offshore oil and gas installations. Sabotage can be defined as "the deliberate destruction, disruption or damage of equipment" by dissatisfied employees. The threat of internal sabotage comes from "insiders" such as current and former employees of oil companies, contractors, offshore service providers, and other trusted persons affiliated with the offshore oil and gas industry. For the purposes of this analysis, intentional disclosure of sensitive/confidential information to third parties is also considered to be a form of internal sabotage. Malicious actions of disgruntled, dishonest, or terminated employees or other insiders, including collusion between insiders and external adversaries, in carrying out or planning an attack can cause serious disruption to oil and gas installations and operations. There have been several reported instances of participation of insiders in attacks against the oil and gas industry, however, none of those specifically involved offshore installations [18].

8.13 TRAININGS

As offshore operation is prone to hazard, it is advisable that offshore personnel have some basic training in safety and sea survival.

8.14 SUMMARY

- Health, safety, and environmental management is an integral part of any business and is considered to be extremely essential when it comes to managing business in oil and gas sectors. The major risk groups in offshore oil industry are blowouts, hydrocarbon leaks on installations, hydrocarbon leaks from pipelines/risers, and structural failures.
- The first step in all risk assessment or quantitative risk assessment (QRA) studies is hazard identification (HAZID). HAZOP studies are generally carried out to identify potential hazards and operability problems caused by deviations that arise from the design intent.
- Safety measures can be adopted in both design and operational stages to avoid catastrophic incidents. One of the major events that can result in serious consequences in oil and gas industries is fire and explosion. There are many preventive measures for fire and explosion.
- Environmental pollution in the marine ecosystems creates complexities and a variety of emerging problems in environmental management. Environmental management policies are framed by the local and global regulatory authorities, which takes into account the factors of current and future interest.
- Safety saves and safety is everyone's domain. Lack of knowledge can magnify safety hazards at any offshore installation. It is advised to read the safety procedures, operating manuals, and standard work procedures before getting on to actual work. Offshore environments are largely complex, interdependent, and hazardous; hence, your safety and the safety of others is dependent mainly on job knowledge.

REFERENCES

1. E. ONGC, *Offshore Safety Manual*, ONGC.
2. *Human factors*. Available at www.iogp.org/oil-and-gas-safety/human-factors/
3. ASTM, *Standard Practice for Human Engineering Program Requirements for Ships and Marine Systems, Equipment, and Facilities*, American Society for Testing and Materials International (1991).
4. *Tidewater Supply Vessel Heavily Damaged After Hitting Production Platform*. Available at https://gcaptain.com/tidewater-supply-vessel-heavily-damaged-hitting-production-platform/
5. Enggcyclopedia. (2011). *Combustion basics: The Fire Triangle/Fire Tetrahedron*. Retrieved April 24, 2019, from www.enggcyclopedia.com/2011/10/combustion-basics-fire-triangle-tetrahedron
6. *OFFSHORE SYSTEMS*. Available at www.tidelandsignal.com/applications/offshore-systems
7. E. ONGC, *ONGC Sea Survival Training Manual*, ONGC Sea Surviv. Train. Man.

8. B. Vignes and B.S. Aadnøy, Well-integrity issues offshore Norway, *SPE Prod. Oper.* 25 (2010), pp. 145–150.
9. New York Times, *Picture of the day the deepwater horizon explosion*, New York Times, (2012).
10. *Off-Shore Drilling Rigs Eyeball*. Available at https://cryptome.org/2015-info/off-shore-rigs/off-shore-rigs.htm
11. *Life Extension and Assessment of Existing Offshore Structures.* Society of Petroleum Engineers, Doha, Qatar (2011).
12. *671 (16) Safety zones and safety of navigation around offshore installations and structures.* Available at https://puc.overheid.nl/nsi/doc/PUC_1400_14/1/
13. H. Esmaeili, *The Legal Regime of Offshore Oil Rigs in International Law*, Taylor & Francis Group, London (2001).
14. *Safety zones and precautionary areas.* Available at www.maritimenz.govt.nz/commercial/environment/offshore-industry/safety-zones.asp.
15. H. Esmaeili, The protection of offshore oil rigs in international law (Part I), *Aust. Min. Pet. Law J.* 18 (1999), pp. 241–252.
16. *Piper Alpha.* Available at https://en.wikipedia.org/w/index.php?title=Piper_Alpha&oldid=889832453
17. *Oil Rigs, Gas pipelines & Offshore Facilities protection.* Available at www.avista-ventures.com/oilandgas.php
18. M. Kashubsky, Protecting offshore oil and gas installations: security threats and countervailing measures, *J. Energy Secur.* 2007 (2013), pp. 1–7.

9 Legislations and Regulations in Offshore Operations around the World

9.1 INTRODUCTION

In general, the gap between the demand and production of oil and gas and energy has been ever increasing since the 1950s. The global energy demand has been challenging the produced/generated amount ever since. This makes it imperative for the oil and gas industry to look for sources and reservoirs which have not yet been discovered. This led to a new revolution in the oil and gas industry – offshore exploration and production.

According to the United Nations Convention on the Law of the Sea (UNCLOS) [1], countries have the right to water up to several nautical miles from the shore, namely, territorial seas. Beyond that, the waters fall under international waters category and no country has an individual right over them. Majority of the countries have accepted 12 nautical miles as the standard extent for territorial waters [2].

Every country has the right to explore for petroleum and minerals within their territorial water, contiguous water, continental shelf, and exclusive economic zone. Most Indian offshore hydrocarbon operations are outside terrotorial water.

Moreover, each country can rightly set up its own unique set of rules and legislations regarding the health, safety, and environment in offshore oil and gas extraction operations.

With many different rules being common globally, there does exist several differences as well. The differences are obvious due to both regulatory authorities as well as country-wise differences in offshore environments.

It is also worth noting that, as drilling companies move into deeper waters, drilling becomes more complex and costly. In addition, offshore drilling can bear greater risks and hazards such as marine weather and pollution that can be very expensive over time.

One reason for this increased danger is the complex equipment needed to drill at such depths. As offshore drilling continues to be pushed to new depths, with oil companies continuously drilling in deeper waters and penetrating further underground, the technology needed to achieve these feats is extremely complex and not entirely invincible [2].

Equipment and steel are strung out over a long piece of geography starting at the surface and terminating 18,000 ft below the sea floor. Thus, it has many potential weak points.

Another reason for the danger is the harsh offshore environments that pose engineering challenges to offshore drilling equipment. Severe weather, ice, and storms pose risks to the functionality of the rigs. In addition, their distance from land makes it harder for additional rescue personnel to promptly reach these areas in emergency situations [3].

According to Menendez, "The bottom line is that during offshore operations, there is always a risk that not only puts lives on the line, but a risk that puts miles of coastline and the economy on the line as well".

This is not just limited to the Gulf but extends to almost every single offshore facility in the oil and gas industry. As risk is more than one can imagine, rules have to be more stringent.

Thus, it is mandatory to know the laws related to offshore operations which varies from country to country. Hence, we have tried to consolidate the laws of the following major offshore operational countries:

- Europe
- Norway
- Kuwait
- Australia
- Egypt
- Qatar
- Russia
- India
- United States
- Canada
- Saudi Arabia

9.2 EUROPE

The European Parliament and the Council of the European Union passed a treaty on the functioning of European Union related to offshore oil and gas exploration, drilling, and production [4].

The major points covered in the treaty are listed below:

- Article 191 of the Treaty on the Functioning of the European Union establishes the objectives of preserving, protecting, and improving the quality of the environment, and the prudent and rational utilization of natural resources [5].
- The objective of this Directive is to reduce as far as possible the occurrence of major accidents relating to offshore oil and gas operations and to limit their consequences, thus increasing the protection of the marine environment and coastal economies against pollution [6].

Global Offshore Legislations and Regulations

- This Directive should apply not only to future offshore oil and gas installations and operations but, subject to transitional arrangements, also to existing installations.
- Major accidents relating to offshore oil and gas operations are likely to have devastating and irreversible consequences on the marine and coastal environment as well as significant negative impacts on coastal economies.
- Accidents relating to offshore oil and gas operations, in particular the accident in the Gulf of Mexico in 2010, have raised public awareness of the risks involved in offshore oil and gas operations, and have prompted a review of policies aimed at ensuring the safety of such operations. The Commission launched a review of offshore oil and gas operations and expressed its initial views on the safety thereof in its Communication "Facing the challenge of the safety of offshore oil and gas activities" on October 13, 2010 [6,7].
- The risks relating to major offshore oil or gas accidents are significant. By reducing the risk of pollution of offshore waters, this Directive should, therefore, contribute to ensuring the protection of the marine environment and, in particular, to achieving or maintaining good environmental status by 2020 at the latest.
- Offshore oil and gas industries are established in a number of regions of the Union, and there are prospects for new regional developments in offshore waters of Member States, with technological developments allowing for drilling in more challenging environments. Production of offshore oil and gas is a significant element in the security of the Union's energy supply.
- There is a need to clarify that holders of authorizations for offshore oil and gas operations pursuant to Directive 94/22/EC are also the liable "operators" within the meaning of Directive 2004/35/EC of the European Parliament and the Council of April 21, 2004 on environmental liability with regard to the prevention and remedying of environmental damage, and should not delegate their responsibilities in this regard to third parties contracted by them [7].
- While general authorizations pursuant to Directive 94/22/EC guarantee to the licensees exclusive rights for exploring for or producing oil or gas within a given licensed area, offshore oil and gas operations within that area should be subject to continuous expert regulatory oversight by Member States.
- Offshore oil and gas operations should be conducted only by operators appointed by licensees or licensing authorities. The operator can be a third party or the licensee or one of the licensees depending on commercial arrangements or national administrative requirements.
- Operators should reduce the risk of a major accident as low as reasonably practicable to the point where the cost of further risk reduction would be grossly disproportionate to the benefits of such reduction. The reasonable practicability of risk reduction measures should be kept under review in the light of new knowledge and technology developments.
- It is important to ensure that the public is given early and effective opportunity to participate in the decision making relating to operations that can potentially have significant effects on the environment in the Union.

- In accordance with Council Directive 92/91/EEC of November 3, 1992 concerning the minimum requirements for improving the safety and health protection of workers in the mineral-extracting industries through drilling (11th individual Directive within the meaning of Article 16(1) of Directive 89/391/EEC), workers and/or their representatives should be consulted on matters relating to safety and health at work and be allowed to take part in discussions on all questions relating to safety and health at work [8].
- Member States should ensure that the competent authority is legally empowered and adequately resourced to be capable of taking effective, proportionate, and transparent enforcement action, including where appropriate cessation of operations is required, in cases of unsatisfactory safety performance and environmental protection by operators and owners.
- The independence and objectivity of the competent authority should be ensured. In this regard, experience gained from major accidents shows clearly that the organization of administrative competences within a Member State can prevent conflicts of interest by a clear separation between regulatory functions and associated decisions relating to offshore safety and the environment [9].
- However, complete separation of the competent authority from economic development of offshore natural resources may be disproportionate where there is a low level of offshore oil and gas operations in a Member State. In such a case, the Member State concerned would be expected to make the best alternative arrangements to secure the independence and objectivity of the competent authority.
- Specific legislation is needed to address the major hazards relating to the offshore oil and gas industry, specifically in process safety, safe containment of hydrocarbons, structural integrity, prevention of fire and explosion, evacuation, escape and rescue, and limiting environmental effect.
- This Directive should apply without prejudice to any requirements under any other Union legal acts, especially in the field of safety and health of workers at work, in particular, Council Directive 89/391/EEC of June 12, 1989 on the introduction of measures to encourage improvements in the safety and health of workers at work and Directive 92/91/EEC. Environmental impact following a major accident.
- An offshore regime needs to apply both to operations carried out on fixed installations and to those on mobile installations, as well as to the lifecycle of exploration and production activities from design to decommissioning and permanent abandonment.
- The best practices currently available for major accident prevention in offshore oil and gas operations are based on a goal-setting approach and on achieving desirable outcomes through thorough risk assessment and reliable management systems.
- To maintain the effectiveness of major hazard controls in offshore waters of Member States, the report on major hazards should be prepared and, as necessary, amended in respect of any significant aspect of the lifecycle of a production installation, including design, operation, operations when

Global Offshore Legislations and Regulations   251

combined with other installations, relocation of such installation within the offshore waters of the Member State in question, major modifications, and final abandonment.

- Well operations should be undertaken only by an installation technically capable of controlling all the foreseeable hazards at the well location, and in respect of which a report on major hazards has been accepted.
- In addition to using a suitable installation, the operator should prepare a detailed design plan and an operating plan pertinent to the particular circumstances and hazards of each well operation. In accordance with best practices in the Union, the operator should provide for independent expert examination of the well design.
- To ensure safety in design and continuous safe operations, the industry is required to follow the best practices defined in authoritative standards and guidance. Such standards and guidance should be updated based on new knowledge and invention to ensure continuous improvement.
- In view of the complexity of offshore oil and gas operations, the implementation of the best practices by the operators and owners requires a scheme of independent verification of safety and environmental critical elements throughout the lifecycle of the installation, including, in the case of production installations, the design stage.
- In so far as mobile offshore drilling units are in transit and are to be considered as ships, they are subject to international maritime conventions, in particular, SOLAS, MARPOL, or the equivalent standards of the applicable version of the Code for the construction and equipment of mobile offshore drilling units (MODU Code).
- The report on major hazards should, inter alia, take into account risks to the environment, including the impact of climatic conditions and climate change on the long-term resilience of the installations. Given that offshore oil and gas operations in one Member State can have significant adverse environmental effects in another Member State, it is necessary to establish and apply specific provisions in accordance with the UN/ECE Convention on Environmental Impact Assessment in a Trans boundary Context done at Espoo (Finland), on February 25, 1991 [10].
- Operators should notify Member States without delay if a major accident occurs, or may be about to occur, so that the Member State can initiate a response, as appropriate.
- To ensure effective response to emergencies, operators should prepare internal emergency response plans that are site-specific and based on risks and hazard scenarios identified in the report on major hazards, submit them to their competent authority, and maintain such resources as are necessary for the prompt execution of those plans, when needed.
- Best global practice requires licensees, operators, and owners to take primary responsibility for controlling the risks they create by their operations, including operations conducted by contractors on their behalf, and therefore to establish, within a corporate major accident prevention policy, the mechanisms and highest level of corporate ownership to implement that policy.

252 Offshore Operations and Engineering

- Responsible operators and owners should be expected to conduct their operations worldwide in accordance with best practices and standards. Consistent application of such best practices and standards should become mandatory within the Union.
- While recognizing that it may not be possible to enforce application of the corporate major accident prevention policy outside of the Union, Member States should ensure that operators and owners include their offshore oil and gas operations outside of the Union in their corporate major accident prevention policy documents.
- Information on major accidents in offshore oil and gas operations outside the Union can help in further understanding their potential causes in promoting learning key lessons and in further developing the regulatory framework.
- Member States should expect operators and owners, in following best practices, to establish effective cooperative relationships with the competent authority, supporting best regulatory practice by the competent authority and to proactively ensure the highest levels of safety, including, where necessary, suspending operations, without the competent authority needing to intervene [1,11].

9.3 NORWAY

Since the beginning of the petroleum exploration and production on the Norwegian Continental Shelf, the need for safe operations had been pointed out, both by the industry and the state authorities. In an industry like this, the protection of people and the environment always stays at the forefront. Continuously improving the level of safety in offshore petroleum operations aims to avoid accidents which would cause significant harm to health, environment, and installations [1,12].

As a starting point, the use of advanced technological materials on offshore and onshore installations plays a very important role in safety – the more recent technological developments are used, the higher are the levels of safety. Therefore, safety regulations oblige operators/licensees to continuously upgrade their equipment standards, often following the advice of experienced scientists. They have to remain updated with the latest developments and incorporate them into the industry.

As the industry is organized in a rather complicated manner, with several different people and companies involved, other important factors should also be taken into consideration.

For all the above reasons, it is evident that a set of political, organizational, economic, and even psychological aspects intervene and determine safety in offshore activities. All these should be taken into account when trying to reduce the possibility of an accident [1].

9.3.1 APPLICABLE LEGISLATION

- The most important and basic legislation is the Royal Decree which formed the Regulations relating to health, safety, and the environment in petroleum activities and in certain onshore facilities (the Framework Regulations).

Global Offshore Legislations and Regulations

These Regulations give the basic rules in offshore petroleum activities. Pursuant to them, there are four supplementary Regulations relating to Management, Facilities, Activities, and Technical and Operational matters that complete the picture of safety [1,13].

- The Act which is widely applicable on offshore operations is the Petroleum Act and describes a variety of oil activities, stipulating that "the petroleum activities shall be conducted in such manner as to enable a high level of safety to be maintained and further developed in accordance with the technological development". This constitutes the basic and general safety requirement in petroleum operations in Norway.
- The Working Environment Act of 2005 regulates a variety of working conditions such as working hours and how working life shall be protected. Therefore, it also applies to offshore personnel [1, 14].
- In addition to this, the Ship Safety and Security Act of 2007 safeguards "life, health, property and the environment by facilitating a high level of ship safety and safety management ensuring a good working environment and safe working conditions on board ships". Therefore, this Act is applicable for personnel resting on board a ship or in a floating device, which usually stands by the oil installations to carry out various supporting tasks [1, 15].

9.4 KUWAIT

The law has been promulgated by the state of Kuwait to assure:

- Maximum ultimate recovery of its hydrocarbon resources
- To prevent waste or pollution
- To achieve safe and efficient practices
- To acquire useful information regarding petroleum operations

The Regulations are set out for general application within the state of Kuwait and those land and offshore areas under the jurisdiction and administration of the state of Kuwait. The Regulations shall apply to every well drilled within the state of Kuwait and to any product obtained or obtainable therefrom; to every production or injection operation; to every petroleum processing and refining operations and the productions derived therefrom; and to the transportation and marketing of crude oil, natural gas, and products derived therefrom. These Regulations shall be implemented by the Technical Affairs (TA) of the Ministry of Oil of the state of Kuwait.

9.4.1 ARTICLE 1

In the implementation of the provisions of this law, the term "Petroleum" shall mean all natural hydrocarbons, whether in solid, liquid, or gaseous state, which are or can be produced from the surface of the ground or from underground and all hydrocarbons or other kinds of fuels derived therefrom.

The term "Petroleum operation" or "Petroleum operations" shall mean reconnaissance and exploration for petroleum; the development of fields, the drilling of wells; the production of petroleum, its treatment, refining, processing, storing, transporting, loading, and shipping; constructing, installing, and operating facilities for water, power, housing, and camps or any other facilities or installations or equipment necessary for accomplishing the aforementioned purposes, and all administrative activities relating thereto.

The term "Authorized operator" shall mean any person, either an individual or a corporate body, legally authorized to carry out any petroleum operation.

9.4.2 ARTICLE 2

All petroleum operations shall be subject to the provisions of this law and its implementing regulations.

Petroleum operations must be carried out in the best manner using efficient and reasonable methods and good techniques as would be expected from a person fully experienced in such operations under similar circumstances and conditions.

9.4.3 ARTICLE 3

Authorized operators shall take all measures and precautions necessary to prevent any damage or danger which might result from petroleum to human life, public health, properties, natural resources, cemeteries or archaeological, religious and tourist sites. They must also take all necessary precautions to prevent pollution of air and surface and underground waters.

9.4.4 ARTICLE 4

All machinery, equipment, and materials used in petroleum operations must conform to recognized internationally acceptable specifications, meet safety requirements, and serve its purpose in accordance with the best techniques in practice.

9.4.5 ARTICLE 5

Authorized operators shall submit periodically to the Minister of Finance and Oil the programs, reports, statements, and information relating to their petroleum operations. Officials designated by the Minister of Finance and Oil shall have the right to inspect the offices of an authorized operator and the sites of his operations, and to examine his documents and records to ascertain that the reports' statements and information submitted under the preceding paragraph are in conformity with the facts.

9.4.6 ARTICLE 6

Before starting to carry out any project relating to petroleum operations, an authorized operator shall submit to the Minister of Finance and Oil a description of the project, including the plans relating thereto its location, capacity, estimated costs,

Global Offshore Legislations and Regulations 255

the operating methods to be used, engineering data and any other information. The Minister shall have the right to approve the project, reject it, or request that it be further studied, clarified, or modified.

In the event of approval of the project, the authorized operator must notify the Minister of the completion of each of its stages to enable the Minister to ascertain that each stage has been carried out in accordance with the description and plans on the basis of which the project was approved.

9.4.7 Article 7

The Regulations necessary for implementing this law shall be issued by the Minister of Finance and Oil. These Regulations shall cover all aspects of petroleum operations with a view to ensuring the exploitation of petroleum resources in accordance with sound practices and efficient methods, the prevention of loss or waste of such resources, and the maximum possible yield therefrom. These Regulations shall also include the necessary measures for regulating the production of petroleum, provisions for safety precautions, and cover statements, information, and documents, which the authorized operator is required to Submit. The procedures to be adopted in this respect and the methods and procedures for carrying out all petroleum operations include the following:

- All activities relating to reconnaissance and exploration operations.
- All activities relating to drilling operations, including permission to drill, well spacing and location, equipping of well, electrical logging, coring and testing, plugging, use, abandonment or suspension of wells, and completion and re-completion of wells.
- All activities relating to production operations including well performance testing and in-hole surveys, workover and maintenance of wells, reservoir behavior studies, enhanced recovery projects, underground storage of petroleum, separation of gas from oil, utilization of gas, and disposal of salt water and reservoir unitization. In relation to the carrying out of these activities, the Minister of Finance and Oil may notify an authorized operator of the actions to be taken and specify for him a time limit for taking them. If the authorized operator does not comply with such notice or if he has complied but failed to achieve the required effect, the Minister of Finance and Oil may issue an order to shutdown production from one well, several wells, or from the reservoir.
- All activities relating to the treatment of petroleum, its refining, processing, storing, transporting, loading, and shipping, which activities include measurements, Calibration and laboratory analyses and their evaluation.
- All activities relating to installations, facilities, and equipment for all petroleum operations.

These regulations shall also cover the measures necessary for the implementation of Article 3 of this law.

9.4.8 ARTICLE 8

An authorized operator shall render, free of charge, to the officials of the Ministry of Finance and Oil who are designated by the Minister of Finance and Oil to implement the provisions of this law, all necessary services including furnished offices, suitable living accommodation, means of transportation, necessary facilities to conduct laboratory tests, and any other necessary services.

9.4.9 ARTICLE 9

The Minister of Finance and Oil may, in certain special cases, entrust any person, individual, or corporate, with the exercise of all or some of the powers vested in him under the provisions of this law and its implementing regulations. Whoever is so entrusted shall have the same rights and privileges as are granted to the officials of the Ministry of Finance and Oil designated by the Minister to implement the provisions of the law and its regulations.

9.4.10 ARTICLE 10

An administrative penalty of not <10,000 dinars shall be impost for a contravention of the provisions of this law or its implementing regulations. In the event of a similar contravention being committed within 3 years from the date of the previous contravention, the administrative penalty shall be doubled. Appropriate compensation shall be imposed in addition to the aforementioned penalty in all cases in which the contravention results in damage to petroleum resources. The imposition of an administrative penalty or compensation shall have no effect on any other sanctions or penalties mentioned in any other laws or regulations or provided for in contractual or international agreements.

9.4.11 ARTICLE 11

An administrative penalty and compensation shall be imposed by a reasoned Decision of the Minister of Finance and Oil on the basis of a report by the competent administrative authority. The Decision shall contain:

- A description of the contravention.
- The provision that is contravened.
- The person responsible for the contravention.
- The amount of the administrative penalty and compensation, if any.

The authorized operator shall be notified of the Decision within 1 week of the issue thereof by a letter by registered post with acknowledgement.

The authorized operator may appeal against this Decision within 21 days from the date on which he is notified thereof, and the appeal shall be made by a written submission to the High Court of Appeal. The Decision shall become enforceable if it is not appealed against, or if it is appealed against and is confirmed, or if the Minister

Global Offshore Legislations and Regulations

issues a decision for the temporary enforcement thereof pending a decision on the appeal. In cases referred to in the preceding paragraph, the Decision shall have the authority of a court writ and shall be enforced by the same procedures followed in the enforcement of court judgments.

9.4.12 ARTICLE 12

The Ministers, each within their jurisdiction, shall implement this law, which shall come into effect on the date of its publication in the official gazette [16,17].

9.5 AUSTRALIA

Offshore Exploration and Production (E&P) to maintain uniformity activities in oil and gas industry involve various activities including lease from government, exploration activities, petroleum production licenses, infrastructure licenses, pipeline licenses, and renewal of permits along with standard procedures in each stage of E&P activities. With regard to legislative aspects, these activities incorporate petroleum prospecting and access authorities; variation, suspension, and exemption of permits; Surrender and cancellation of titles; and other provisions. Australian Commonwealth government has compiled legislation related to offshore E&P activities, which incorporates following acts and amendment acts.

- Offshore Petroleum and Greenhouse Gas Storage Act 2006
  - This act incorporates the agreement between the Commonwealth and State, simplified maps describing offshore areas of states and territories, joint authorities for different states and territories, royalty, petroleum exploration permits, petroleum retention permits, petroleum production licenses, infrastructure license, pipeline license, standard procedures, exemption, title related decisions, jurisdiction of courts, liability for act omissions, personal property securities, Commonwealth reserves, scheduled area of states and territories, and occupational health and safety [18].
  - Additionally, National Offshore Petroleum Safety and Environmental Management Authority (NOPSEMA) has a power to inspect companies. Petroleum environmental inspection is carried out by NOPSEMA to monitor compliance with petroleum environmental laws.
- Offshore Petroleum (Royalty) Act 2006
  - Main aspects of this act are imposition of royalty in different, rate of royalty, reduction, and exemption in royalty. Rate of royalty without any exemption is 10% of the production at wellhead [19].
- Offshore Petroleum and Greenhouse Gas Storage (Regulatory Levies) Act 2003
  - In this act, levies for various conditions are described in reference with regulation which are in force. Levies include safety investigation levy, safety case levy, well investigation levy, annual well levy, well activity levy, annual titles administration levy, and environmental plan levy.

258　　Offshore Operations and Engineering

- Offshore Petroleum and Greenhouse Gas Storage (Registration Fees) Act 2006
  - The OPGGS (Registration Fees) Act, 2006 describes fees imposed on titleholder to transfer the title. It also gives insight about dealings done in reference to petroleum title.
- Offshore Petroleum (Repeals and Consequential Amendments) Act 2006
- Petroleum (Timor Sea Treaty) Act 2003
- The Ministerial Council, the Joint Commission, and the Designated Authority exercise the rights and responsibilities of Australia, in relation to the exploration, development, and exploitation of the petroleum resources of the JPDA in accordance with this Treaty.
- Petroleum (Timor Sea Treaty) (Consequential Amendments) Act 2003
- The amendments done under this act include Crimes at Sea Act 2000, Customs Act 1901, Fringe Benefits Tax Assessment Act 1986, Income Tax Assessment Act 1936, International Organizations (Privileges and Immunities) Act 1963, Migration Act 1958, Passenger Movement Charge Collection Act 1978, Petroleum (Submerged Lands) Act 1967, Petroleum (Timor Gap Zone of Cooperation) Act 1990, Quarantine Act 1908, Superannuation Guarantee (Administration) Act 1992, Taxation Administration Act 1953, and Workplace Relations Act 1996 [20].
- Offshore Petroleum Amendment (Greater Sunrise) Act 2007
- Petroleum (Submerged Lands) Legislation Amendment Act 2001

The regulations enforced by federal legislation, which regulates the offshore activities related to oil and gas industry, are listed below.

- Offshore Petroleum and Greenhouse Gas Storage (Safety) Regulations 2009
- Offshore Petroleum and Greenhouse Gas Storage (Resource Management and Administration) Regulations 2011
- Offshore Petroleum and Greenhouse Gas Storage (Environment) Regulations 2009
- Offshore Petroleum and Greenhouse Gas Storage (Regulatory Levies) Regulations 2004 [21].

Titleholders have to follow these regulations to form title condition. Breach of this leads to compliance action by federal government. However, titleholders can apply for exemption or variation of a title under section 264 of the OPGGSA, which is judged by the Joint authority and Commonwealth.

The National Offshore Petroleum Titles Administrator (NOPTA) and the National Offshore Petroleum Safety and Environmental Management Authority (NOPSEMA) are bodies that have reviewed, and revised, the administrative guidelines for offshore petroleum activities. The required changes are included to ensure the ease of new-comers to acquaint with the Joint authority. The updates also bring the guidelines in line with the current Joint Authority expectations in relation to the content of applications for acreage release areas, exploration permit renewals, work programs, the

Global Offshore Legislations and Regulations

declaration of locations, petroleum retention leases, petroleum production and infrastructure licenses, and pipeline licenses. The designated authorities were abolished to carry out such work by NOPTA from January 1, 2012. NOPTA also takes care of environmental integrity with industry [22].

States and territories have included some acts related to offshore activities in their legislation, which are listed below.

- Petroleum (Offshore) Act 1982—New South Wales

  An Act relating to the exploration for, and the exploitation of, the petroleum resources, and certain other resources, of certain submerged lands adjacent to the coasts of New South Wales; to repeal the Petroleum (Submerged Lands) Act 1967 and the Petroleum (Submerged Lands) Taxation Act 1967; to amend the Petroleum Act 1955 and the Pipelines Act 1967; and for other purposes.

- Petroleum (Submerged Lands) Act 1981—Northern Territory

  The Act was made for provision with respect to the exploration for, and the exploitation of, petroleum resources, and certain other resources, of certain submerged lands adjacent to the coasts of the Northern Territory, and for other purposes.

- Petroleum (Submerged Lands) Act 1982—Queensland

  An Act to make provision with respect to the exploration for, and the exploitation of, the petroleum resources, and certain other resources of, and to convey petroleum resources (wherever recovered) across certain submerged lands adjacent to the coasts of Queensland, and for other purposes.

- Petroleum (Submerged Lands) Act 1982—South Australia

  An Act to make provision with respect to the exploration for, and the exploitation of, the petroleum resources, and certain other resources, of certain submerged lands adjacent to the coasts of South Australia, and for other purposes.

- Petroleum (Submerged Lands) Act 1982—Tasmania

  An Act to make provision with respect to the exploration for, and the exploitation of, the petroleum resources, and certain other resources, of certain submerged lands adjacent to the coast of the State of Tasmania, and to provide for related matters (Royal Assent June 30, 1982) Preamble.

- Offshore Petroleum and Greenhouse Gas Storage Act 2010—Victoria

  The purpose of this Act is to (a) re-enact (with modifications) provisions regulating petroleum exploration and recovery activities and petroleum facilities; and (b) provide for the regulation of geological storage of carbon dioxide in the Victorian offshore area.

- Petroleum (Submerged Lands) Act 1982—Western Australia

  An Act to make provision with respect to the exploration for, and the exploitation of, the petroleum resources, and certain other resources, of certain submerged lands adjacent to the coast of Western Australia, to repeal the Petroleum (Submerged Lands) Act 1967, and for incidental and other purposes.

260   Offshore Operations and Engineering

- Petroleum (Submerged Lands) Regulations 1990—Western Australia
  The act was made to describe prescribed fees, rate, and sums; from of instrument transfer; royalty value and deducible imposts; and regarding geocentric and geodetic datum.

9.6   EGYPT

Egypt is considered a significant oil and gas producer in North Africa. The Egyptian Ministry of Petroleum is the governmental authority responsible for the regulation and development of the oil and gas industry in Egypt.

The Egyptian Ministry of Petroleum acts mainly through two major entities in the oil and gas fields. The first is the Egyptian General Petroleum Corporation (EGPC), which is a public entity regulating the petroleum industry in Egypt. The second is the Egyptian Natural Gas Holding Company (EGAS), which is a private entity owned by the EGPC responsible for regulating the gas industry in Egypt. The EGPC and EGAS focus on oil and gas activities, adapting an effective action plan to organize and handle the activities of oil and gas resources in Egypt [23].

The EGPC and EGAS are engaged in a wide range of activities, including upstream (exploration and exploitation [drilling and production of oil and gas]) and downstream (processing, transmission, distribution of oil and gas in the domestic market, and marketing thereof).

Under the Egyptian Constitution, all oil and gas resources are under the control of the state. Accordingly, only the state can grant rights for exploration and exploitation of oil and gas resources for interested investors. Rights of exploration and exploitation of the oil and gas resources are granted under the form of a concession agreement. The concession agreement is issued by virtue of a law. The law issuing the concession agreement authorizes the Minister of Petroleum to enter into the concession agreement between Egypt, the EGAS, or EGPC (as the case may be) from one side, and the contractor company willing to undertake the exploration and the exploitation activities from the other side. All concession agreements are published in the Egyptian Official Gazette and generally follow a standard format, with slight deviation in each agreement.

9.6.1   REGULATION

The extraction of oil and gas is regulated by the Egyptian Mining and Quarries Law 86 of 1956 and the terms and conditions set out under the relevant concession agreements.

The Ministry of Petroleum is the sole body with regulatory responsibility for the petroleum sector in Egypt through two principal public companies.

- Egyptian General Petroleum Company (EGPC).
- Egyptian Natural Gas Holding Company (EGAS).

Permission must be obtained first from the Ministry of Petroleum by the entity proposing to extract oil or gas. Permission usually takes the form of a concession agreement.

Global Offshore Legislations and Regulations

Certain principal laws regulate the oil and gas sector, and include:

- Law No. 86 of 1958 organizing Mines and Quarries and executive regulations.
- Law No. 20 of 1987 organizing the Egyptian General Petroleum Company (EGPC).
- Law No. 217 of 1980 organizing Natural Gas and executive regulations.
- Law No. 4 of 1988 regarding Oil Pipelines.

Regulation is mainly achieved by standard terms and provisions in the concession agreement signed by the Egyptian Minister of Petroleum and the Contractor. However, projects can differ from case to case [24].

EGAS is managed by a board of directors appointed by its General Assembly on the advice of the Minister of Petroleum. The Chairman of the General Assembly is the Minister of Petroleum.

In practice, EGAS has two major responsibilities:

- Exclusive off-taker of gas for the local market.
- Party beneficiary of the concession agreements for all new gas concessions. It receives the government's production share, which is usually 70%–80% of revenue after the deduction of costs by the contractor.

9.6.2 DIRECTIVE ON OFFSHORE SAFETY

The key objectives of the Directive are:

- To reduce the occurrence of major accidents relating to offshore oil and gas operations;
- To establish a framework for the safe exploration and production of oil and gas, thereby increasing the protection of the marine environments against pollution;
- If prevention fails, to ensure that clean-up and mitigation are carried out to limit the consequences; and
- To improve the response in the event of an incident.

9.7 QATAR

Qatar is a major hydrocarbon province and a net exporter of oil, gas, liquid fuels, and petroleum products. Qatar's reign as a major oil producer began in 1935 with the discovery of the Dukhan onshore field. It has been a member of the Organization of the Petroleum Exporter Countries (OPEC) since 1961. The country ranks 13th in the world for crude oil reserves, estimated as of January 2014 at 25.2 billion barrels [25].

In the past few decades, Qatar's economy has shifted from being based primarily on production of conventional oil to production of natural gas and non-crude liquids. Qatar holds about 13% of the world's conventional reserves of natural gas, placing it third in the global rankings, with estimated reserves as of January 2014 at 885 trillion ft^3. Most of Qatar's natural gas reserves are located in the giant offshore North Field, which spans an area that is roughly equivalent to Qatar itself. The North Field is considered the world's largest non-associated natural gas field, and traverses Qatar's and Iran's maritime territories.

In terms of production, Qatar is currently ranked as the world's fourth largest dry natural gas producer (after the United States, Russia, and Iran). Gas production in 2012 was 5.5 trillion cf. According to the recent statements by the Minister of Energy and Industry, Qatar's natural gas production exceeded 7 trillion ft^3 in 2013.

9.7.1 GOVERNMENT POLICY OBJECTIVES

The Qatar National Development Strategy 2011–2016 (Development Strategy) is the first national development strategy for Qatar prepared within the framework of Qatar's National Vision 2030. The plan affirms the state's commitment to responsible exploitation of Qatar's hydrocarbon resources, and the judicious investment of the proceeds for the benefit of current and future generations. It pledges to continue investments in infrastructure, people, and institutions, and to implement a concerted drive toward greater economic efficiency and improved competitiveness for advances in productivity and economic diversification [26].

Specifically, the Development Strategy outlines plans for investments by Qatar Petroleum and related companies of about QAR88 billion from 2011 to 2016, including QAR7 billion in expanded capacity of Qatar companies in the petrochemical sector, including producers of low-density polyethylene, ammonia, and urea. The Government also plans more than US$65 billion in infrastructure spending through 2016, including improvements in the power and water sectors and a new Doha port.

9.7.2 REGULATION

The Ministry of Energy and Industry regulates Qatar's oil and natural gas policy, which is subject to the ultimate control of the Emir of Qatar. Under the Qatar Petroleum Law, Qatar Petroleum manages upstream, midstream, and downstream oil and gas operations on behalf of the Government. Qatar Petroleum acts as the state's investment arm in the oil and gas sector [27].

9.7.3 THE REGULATORY REGIME

The oil and gas industry (including LNG) is overall regulated by the Natural Resources Law (Law No. 3) of 2007 regarding the Exploitation of Natural Resources and its Sources (Natural Resources Law). Oil and natural gas, as well as other mineral resources, are the property of the state. Qatar Petroleum is entrusted with management and development of all of Qatar's hydrocarbon resources. There is little detail in the Natural Resources Law as to how this is to be implemented.

The right to explore, develop, and produce petroleum is typically granted by way of development and production sharing agreements (DPSAs), and exploration and

Global Offshore Legislations and Regulations 263

production sharing agreements (EPSAs) with Qatar Petroleum. Qatar Petroleum (and, through it, the Government), therefore, determines the basis on which an entity participates in the Qatari oil and gas (including LNG) industry. LNG projects are usually integrated with the corresponding upstream development. The key legislation in relation to oil and gas exploration and production include:

- Decree-Law No. (4) Of 1977 on Preserving Oil Wealth.
- Decree-Law No. (10) Of 1974 concerning the Establishment of Qatar Petroleum (and its amendments).
- Decree-Law No. (30) Of 2002 issuing the Environmental Protection Law.
- Law No. (8) Of 2004 concerning Protection of the Maritime Facilities of Petrol and Gas.
- Decision No. (4) of 2005 of the President of the Supreme Council of Environment and Natural Protection concerning the issuance of the Executive Regulations of the Environmental Protection Law issued by Decree-Law No. (30) Of 2002.
- Law No. (3) Of 2007 regarding the Exploitation of Natural Resources and its Sources.
- Decree- Law No. (15) Of 2007 concerning the Organization of Marketing and Sale of Regulated Products outside the State of Qatar (Tasweeq Law) [28].
- Rights to oil and gas [27].

9.8 RUSSIA

Russia is one of the world's leading producers and exporters of both oil and gas. Its proven oil reserves total approximately 109.5 billion barrels, which equates to nearly 6.4% of the total global reserves.

The Russian domestic oil and gas sector has been facing more challenging times in recent years due to low oil prices and limited access to financing due to international sanctions. The state policy for oil and gas has been focused on maintaining the current production levels and supporting state companies.

9.8.1 REGULATORY BODIES

9.8.1.1 Oil and Natural Gas

The licensing regime is administered by the Ministry of Natural Resources and Ecology of the Russian Federation and federal agencies under its jurisdiction. Subordinate to that Ministry, the Federal Agency for Subsoil Use (*Rosnedra*) is the administrative agency primarily responsible for the regulation of oil and gas extraction. Rosnedra is responsible for:

- Issuing, suspending, and revoking subsoil use licenses.
- Approving deposit development plans.
- Transferring and storing geological information.

The Federal Service for Supervision of Nature Use (*Rosprirodnadzor*) oversees compliance with legislation regulating subsoil use and protection of the environment. Additionally, the Federal Environmental, Industrial and Nuclear Supervision Service (*Rostekhnadzor*) issues mining allotments that determine deposit boundaries, safety certificates, and operating licenses [29].

There are two kinds of regimes provided for under Russian law:

- **Tax-royalty regime.** This is, by far, the dominant regime governing oil and gas extraction in Russia. The main body of legislation is contained in the Federal Law on Subsoil dated February 21, 1992 (Subsoil Law), and relevant regulations.
- **Production sharing regime.** The provisions governing production sharing regimes are set out in the Federal Law on Production Sharing Agreements dated December 30, 1995 (PSA Law), which has been substantially amended. All existing production sharing agreements in Russia, however, were concluded before the entry into force of the PSA Law, and therefore predate many of its provisions.

Normally, subsoil licenses terminate on expiration of their designated term, but they can also be revoked by state authorities before expiration of their term for the following reasons:

- Appearance of immediate danger to the health of the people working or living in the areas affected by operations related to subsoil use.
- Violation by the subsoil user of material terms of the license.
- Systematic violation by the subsoil user of the established rules for subsoil use.
- Occurrence of emergency situations (natural disasters, war, and others).
- Subsoil user's failure to commence operations in accordance with the established scope and term of the license.
- Liquidation of an enterprise or other subject of economic activities that holds the license for subsoil use.
- Subsoil user's failure to file the reports required by Russian law.
- At the initiative of a subsoil user on submission of the appropriate application.

In situations involving the appearance of immediate danger or occurrence of emergency situations, the subsoil use rights are terminated immediately after the authorities decide that it is necessary, provided that the subsoil user has been served a written notice of the decision. Where there is violation of material terms, systematic violations or failure to commence operations within the established scope, the subsoil use rights can be terminated if the subsoil user fails to remedy the violations within 3 months of receiving a written notice of them.

When subsoil use rights are terminated, the subsoil user must decommission the operations in the field at its own cost.

Global Offshore Legislations and Regulations 265

9.8.2 RUSSIAN LEGISLATION REGULATING FOREIGN INVESTMENTS

In July 2017, Russian legislation regulating foreign investments was modified through two sets of amendments. The first set of amendments relates to the limitations of so-called "offshore companies" when investing in Russian strategic companies and participating in the privatization of Russian state assets. The second set of amendments extended the application of the Law "On Foreign Investments" to certain Russian entities, which means that the Governmental Commission for Control over Foreign Investments (Governmental Commission) will have wider control over investments made in "strategic companies" [30].

9.8.3 RULES FOR OFFSHORE COMPANIES

On July 1, 2017 the following amendments to the Law "On Foreign Investments in Strategic Companies" and the Privatisation Law were adopted and entered into force with immediate effect:

- Investments made by so-called "offshore companies" in Russian strategic companies are now treated in the same way as foreign states or international organizations that invest in Russian strategic companies, meaning that such investments are subject to stricter rules. Before the amendments, investments through "offshore companies" in strategic companies were not specifically regulated and subject to ordinary foreign strategic investment control mechanism applicable to any other foreign investor.
- "Offshore companies" are companies registered in particular jurisdictions or offshore zones, a list of which has been approved by the Ministry of Finance of the Russian Federation. At present, offshore zones include 40 territories, including the British Virgin Islands, the United Arab Emirates, the Principality of Monaco, Gibraltar, the Special Administrative District of Hong Kong (Xianggang), and others. Typical foreign holding jurisdictions for Russian investments, such as Cyprus, Luxembourg, or the Netherlands, are not included in the list, and therefore, companies incorporated in these jurisdictions are not considered "offshore companies".
- Restrictions in relation to offshore companies also extend to companies controlled by offshore companies, including Russian companies.
- The following restrictions for "offshore companies" have been introduced with the recent legislative amendments:
- An unconditional prohibition on "offshore companies" (and entities controlled by an "offshore company") acquiring more than 50% of the shares in a strategic company or otherwise acquiring control over a strategic company.
- An unconditional prohibition on "offshore companies" (and on entities controlled by an "offshore company") acquiring 25% or more of the shares in a strategic mining company operating subsoil plots of federal importance (subsoil user).

266

- An unconditional prohibition on "offshore companies" (and on entities controlled by an "offshore company") acquiring 25% or more of the book value of the main production assets of a strategic company.
- "Offshore companies" (or entities controlled by an offshore company or a group of persons which includes an offshore company) will no longer be able to act as buyers of state-owned or municipal-owned property.
- The acquisition by an "offshore company" (or an entity controlled by an "offshore company") of more than 25% of the shares in a strategic company now requires approval by the Governmental Commission (the same applies to the acquisition of other rights to block decisions of strategic companies).
- The acquisition by an "offshore company" (or an entity controlled by an "offshore company") of more than 5% of the shares in a strategic company that is a subsoil user now requires approval by the Governmental Commission (the same applies to the acquisition of other rights to block the decisions of strategic subsoil user companies) [31].

These rules do not extend to transactions of "offshore companies", where the beneficial owner is a citizen and a tax resident of the Russian Federation, or where the Russian Federation itself or its constituent entity is the ultimate owner of the "offshore entity".

9.9 INDIA

The oil and gas sector is one of the core industries in India and contributes approximately 15% to its total gross domestic product (GDP). The industry was approximately worth US$139.8 billion in 2015, and has tremendous growth potential owing to recent government policies aimed at increasing highway and road infrastructure, promoting Indian manufacturing, creating dedicated freight corridors, establishing smart cities, and so on [32].

9.9.1 DOMESTIC PRODUCTION

In 2015, India was the fourth largest consumer of crude oil and petroleum products after the United States, China, and Japan. Despite having good fossil fuel resources, India increasingly relies on crude oil imports to meet its domestic energy demands. Currently, about 80% of India's crude oil is imported from overseas.

India produced 41.2 million tonnes of crude oil in 2015 (0.9 million barrels a day). The share of offshore crude oil production for 2015 was around 50.2% and the balance was from six onshore states:

- Andhra Pradesh
- Arunachal Pradesh
- Assam

Global Offshore Legislations and Regulations

- Gujarat
- Rajasthan
- Tamil Nadu

National oil companies account for 70% of the total oil produced in India, with the remaining 30% coming from private/joint venture companies.

9.9.2 GOVERNMENT POLICY OBJECTIVES

The Integrated Energy Policy was adopted by the government in 2008, providing a collective policy regime covering all sources of energy. The broad vision behind this policy has been to create a regime to reduce dependence on imports and reliably meet energy demands with safe, clean, and convenient energy at minimum cost. To supplement this, the government and several states have adopted policies promoting clean and renewable energy.

9.9.3 REGULATION

The Ministry of Petroleum and Natural Gas manages and oversees upstream oil and natural gas exploration and production.

The Directorate General of Hydrocarbons is the agency vested with the responsibility of promoting sound management of Indian petroleum and natural gas resources with due regard to environmental, safety, technological, and economical aspects of petroleum activities.

India has a quasi-federal constitution where both the federal and the state governments have legislative powers. However, under the Indian constitution, only the federal government is empowered to make laws relating to regulation and development of oil fields and mineral oil resources, petroleum, and petroleum products.

From time to time, the government formulates policies under which concessions for exploration of oil and gas are awarded through a transparent competitive bidding system to private/foreign investors and national oil companies on the same fiscal and contractual terms [33].

9.9.4 LEGAL FRAMEWORK ON MINERALS MINING IN INDIA

The Mines and Minerals (Development & Regulation) Act (MMDR), 1957 is the principal legislation that governs the mineral and mining sector in India. The Act is a central legislation in force for regulation of mining operations in India. Under the act, minerals are grouped under two broad heads, major minerals and minor minerals. The list is lucid [34].

The power to frame policy and legislation on minor minerals are entirely subjected and delegated to the state governments while policy and legislation relating to the major minerals are dealt by the Ministry of Mines under the Union/Central Government of India. The central government has the power to notify "minor minerals" under Section 3 (e) of the MMDR Act, 1957. On the other hand, as per Section 15 of the MMDR Act, 1957, state governments have complete powers for making Rules for grant of concessions in respect of extraction of minor minerals, as well as levy and collection of royalty on minor minerals.

Whereas in case of offshore areas (territorial waters, continental shelf, exclusive economic zone, and other maritime zones of India), the ownership of minerals vests exclusively with the Central Government. To regulate the mining and development of minerals in the offshore area, the Parliament has enacted the "Offshore Areas Minerals (Development and Regulation Act, 2002". The Act empowers the Central Government to grant mineral concessions for offshore areas and collect royalty. The Indian Bureau of Mines has been notified as the administrative authority for concession management of offshore areas.

9.9.5 Offshore Areas Minerals (Development & Regulation) Act, 2002

The Act is applicable to all minerals in offshore areas including minerals prescribed under the Atomic Energy Act, 1962, but excludes oils and related hydrocarbons as there is separate legislation for them in force. The Act came into effect from 15.1.2010 vide S.O.338 (E), dated 11.2.2010 notified by the Central Government.

Indian Bureau of Mines has been notified as the "administering authority" and "authorized officer" under Section 4 and Clause (i) of Section 22 of the Act vide S.O.339 (E) and 340(E) dated 11.2.2010. The Secretary, Ministry of Mines has been notified as the "Authorized Officer" to hear and decide cases relating to Clauses (a) and (b) of Section 28(1) vide S.O.341 (E) dated 11.2.2010.

The Act empowers the Central Government to make rules for the purpose of the Act including terms and conditions under the reconnaissance permit, exploration license, production lease, etc.

The Government of India announced the New Exploration Licensing Policy (NELP) in 2000 under which blocks for exploration of oil and gas were on offer for bidding. The NELP provides an international class fiscal and contract framework for exploration and production of hydrocarbons [35].

9.9.6 Offshore Areas Mineral Concession Rules, 2006

1. The Offshore Areas Mineral Concession Rules, 2006 lay down the process for the grant and renewal of reconnaissance permits, exploration licenses and production leases as per provisions of Section 35 of the Offshore Areas Mineral (Development and Regulation) Act, 2002.
2. The rules prescribe measures for protecting the marine environment and safety measures to be followed in the leased area.
3. The rules also define the operational guidelines for each concession granted under the Act [36].

9.9.7 Other Regulatory Requirements

1. Oilfield (Regulation and Development) Act 1948
2. Petroleum and Natural Gas (Safety in Offshore Operations) Rules 2008
3. Oil mines Regulation 1984

Global Offshore Legislations and Regulations 269

9.10 UNITED STATES

The United States is now the largest producer of oil and natural gas in the world. It is also the second-largest producer of liquefied natural gas (LNG), and is expected to become the top producer within the next decade. As a result of increasing production rates driven by technological advances in onshore horizontal drilling and high-volume hydraulic fracturing, the United States now produces nearly 10,000,000 barrels per day (Bpd) of crude oil, up from an average of just over 5,000,000 Bpd during 2005–2010. The oil and natural gas production sector is a staple of the US economy, employing approximately 200,000 people in 2014 [37].

9.10.1 GOVERNMENT POLICY OBJECTIVES

The US government does not have a national energy policy. However, the oil and gas industry can be affected by tangential government energy and environmental policies, such as automotive fuel efficiency standards.

Individual states within the United States have developed specific policy objectives, most commonly stated as a policy to prevent waste and protect the environment while promoting the greatest ultimate recovery of indigenous oil and gas from within the state.

The United States has both federal and individual state agencies that regulate certain aspects of oil and gas production. Neither the US federal government nor the individual states have established a comprehensive energy policy to manage their energy resources. For example, domestic onshore oil and gas development is regulated by individual states under mandates to prevent waste and protect human health and the environment, while encouraging the greatest ultimate use of domestic oil and gas production. Oil and gas production occurring offshore in the Gulf of Mexico is managed by various US federal government agencies to ensure safe and environmentally responsible development, as well as the payment of production royalties and taxes for the public benefit.

9.10.2 REGULATION

Domestic onshore oil and gas development is regulated by individual states in which the activity will take place. Each state has its own regulatory agency or agencies that control the following:

- The distance between oil wells and property lines to protect the rights of adjacent landowners.
- Prevention of waste.
- Health and safety issues.

However, local government control over oil and gas production is generally not permitted by state law, except for local zoning input that in some states allows local government control over where and when oil and gas production activities can take place (to prevent, for example, residential neighborhoods from noise pollution, industrial traffic, or perceived health hazards).

270 Offshore Operations and Engineering

Individual states also have authority over the taxation of oil and gas production that occurs within the state. Oil and/or gas were produced in 33 of the 50 states within the United States in 2014. Additionally, various agencies of the US federal government regulate oil and gas production in the waters of the Gulf of Mexico, as well as the exploration of oil and gas on federal lands, for example:

- The Department of the Interior regulates the extraction of oil and gas from federal lands.
- The Bureau of Land Management regulates oil development, exploration, and production on federal onshore properties.
- The Office of Natural Resources Revenue collects royalties owed to the government for onshore and offshore production [38].

9.10.3 LEASE/LICENSE/CONCESSION TERM

Oil and gas leases in the United States are usually subject to two separate durations:

- A term of years during which the lessee can explore and develop the property without paying royalties on production.
- If and when oil and/or gas is discovered in producible quantities sufficient to generate royalty payments to the lessor/owner (known as producing in paying quantities), the lease will continue for so long as oil and/or gas is produced from the leased property.

9.11 CANADA

Canada has a set of four principal Acts which govern oil and gas activities in the offshore:

The Canada Petroleum Resources Act governs the lease of federally owned oil and gas rights on "frontier lands" to oil and gas companies that wish to find and produce oil and gas. "Frontier lands" include the "territorial sea" (12 nautical miles beyond the low water mark of the outer coastline), and the "continental shelf" (beyond the territorial sea). It is the statute under which the federal government must first give permission for oil and gas exploration to occur on frontier lands, and provides opportunity for the federal government to protect the environment by attaching exploration restrictions when leasing rights or by stopping work if there is an environmental problem.

Under the Act, subsurface oil and gas rights in unexplored areas are issued during a "public call for bids" and the Minister may attach conditions to the transfer of rights (including conditions for protecting the environment). For each right issued, the successful oil and gas company must pay a royalty to the federal government [39].

The Canada Oil and Gas Operations Act governs the exploration, production, processing, and transportation of oil and gas in marine areas controlled by the federal government. These areas include the "territorial sea" (12 nautical miles beyond the low water mark of the outer coastline), and the "continental shelf" (beyond the

Global Offshore Legislations and Regulations 271

territorial sea). They do not include areas controlled by the provincial government. The purpose of the Act is to promote safety, protection of the environment, the conservation of oil and gas resources, and joint production agreements [40].

The Canada-Newfoundland Atlantic Accord Implementation Act and the Canada-Nova Scotia Offshore Petroleum Resources Accord Implementation Act, otherwise known as the Accord Acts, implement agreements between the federal and provincial governments relating to offshore petroleum resources. The Accord Acts mirror both the COGOA and CPRA and outline the shared management of oil and gas resources in the offshore, revenue sharing, and establishes the respective offshore regulatory boards [41].

9.11.1 REGULATION

The primary legislation governing oil and natural gas activities offshore Newfoundland and Labrador and Nova Scotia is the Atlantic Accord and the Atlantic Accord Implementation Acts. Offshore oil and natural gas operations in Newfoundland and Labrador are regulated by the Canada-Newfoundland and Labrador Offshore Petroleum Board (C-NLOPB), an independent administrative board jointly appointed by the federal and provincial governments, whose mandate is to interpret and apply the provisions of the Atlantic Accord and the Atlantic Accord Implementation Acts to all activities of operators in the Newfoundland and Labrador Offshore Area, as well as to oversee operator compliance with those statutory provisions. Similarly, in Nova Scotia, the Canada-Nova Scotia Offshore Petroleum Board (C-NSOP) regulates the industry [42].

Operators are required to submit a variety of plans and meet specific requirements to receive authorization from the C-NLOPB or C-NSOPB to conduct work offshore. This includes developing and submitting the following:

- A Safety Plan which sets out the procedures, practices, resources, sequence of key safety-related activities, and monitoring measures necessary to ensure the safety of the proposed work or activity;
- An Environmental Protection Plan that sets out the procedures, practices, resources, and monitoring necessary to manage potential hazards and to protect the environment from the proposed work or activity;
- A Contingency Plan (including emergency response procedures such as oil spill response plans) that sets out how to mitigate the effects of any potential event that might compromise safety or environmental protection and includes territorial or federal emergency response plan, and where oil is reasonably expected to be encountered, identify the scope and frequency of the field practice exercise of oil spill countermeasures; and
- A Benefits Plan that describes a plan for the employment of Canadians and, in particular, members of the labor force of the province in which the activity is occurring; and for providing manufacturers, consultants, contractors, and service companies in the province and other parts of Canada with a full and fair opportunity to participate on a competitive basis in the supply of goods and services.

272   Offshore Operations and Engineering

All plans are reviewed by the C-NLOPB or C-NSOPB and are accepted as part of the authorization process prior to initiating the activity. Companies conducting activity offshore additionally follow other federal legislation and regulation including, but not necessarily limited to:

- Canadian Environmental Protection Act
- Canadian Environmental Assessment Act (CEAA)
- Species at Risk Act (SARA)
- Fisheries Act
- Migratory Birds Conservation Act
- Arctic Waters Pollution Prevention Act
- Occupational Health and Safety Transitional Regulations [42].

9.12 SAUDI ARABIA

Offshore oil and gas legislations mainly pertain to the drilling and workover operations as the offshore environment and the rig and personnel are exposed to a majority of risks and hazards during these operations.

These regulations are corporate, regional, national, and global in nature. All operators, like Saudi Aramco, must adhere to these regulations.

Considerations:

- Discharge to marine environment from drilling operations
- Wastewater treatment, reuse, and disposal
- Protection of marine life
- Pollution control and environment protection

Corporate regulations:

Saudi Aramco has strong corporate standards regarding guidelines to environmental protection policy, the major aim being to manage the operations without any adverse effects to the surrounding environment.

The standards are as follows:

- Sewage disposal: A sewage treatment plant is required for any facility located less than 4 nautical miles from land. Moreover, the disposed matter must meet the specified guidelines.
- Industrial drainage: All types of water flows and oil drainage must be collected in slop tanks or caissons.
- Trash/Rubbish: Can't be disposed to the sea and has to be hauled back to an approved onshore disposal site.
- Oil-based mud/toxic fluids/cuttings: These must be hauled back to an approved onshore disposal site.
- Drilled cuttings from OBM: These must be cleaned using the best practical technology, and should be disposed as close to the sea floor as possible.
- Contingency plans: All contractors are required to recognize the company's requirements in the contingency plans related to disaster and oil spill prevention.

Global Offshore Legislations and Regulations 273

- Water-based mud/cuttings: These shall not be discharged if they contains persistent systematic toxins. Toxicity tests (e.g. LC-50) must be run, and appropriate disposal sites must be chosen.
- Ambient air quality: The facilities must comply with the pollutant concentrations mentioned by the Meteorological and Environmental Protection Administration (MEPA).
- Noise: Noise level should not exceed 85 dBa, and personnel should not be exposed to high noise levels for more than a specified time duration.
- Flaring: Flaring must be done only in necessary conditions like in pressure testing and kill operations. Prior permissions must be taken and notified. It should begin in daylight and can continue after sunset as well.

Applicable agreements for drilling operations in the Arabian Gulf

These agreements contain all the material and information related to the safety regulations and specifications offshore in the Arabian Gulf. They come under Regional Organization for Protection of Marine Environment (ROPME).

- ROPME protocol concerning marine pollution resulting from the exploration and exploitation of continental shelf.
- ROPME protocol concerning regional cooperation combating pollution by oil and other harmful substances in case of emergency.
- ROPME protocol for protection of marine environment from land-based sources [43].

REFERENCES

1. Mette Gravdahl-Agerup, *Offshore Safety Regulations the European Perspective*, 933 (2012), pp. 1–62.
2. *TERRITORIAL SEA AND CONTIGUOUS ZONE.* Available at www.un.org/depts/los/convention_agreements/texts/unclos/part2.htm
3. *Why Is Offshore Drilling So Dangerous.* Available at www.livescience.com/32614-why-is-offshore-drilling-so-dangerous-.html
4. *Offshore oil and gas safety.* Available at https://ec.europa.eu/energy/en/topics/energy-security/offshore-oil-and-gas-safety
5. European Commission, *The precautionary principle: decision-making under uncertainty*, Eur. Comm. DG Environ. by Sci. Commun. Unit (2017), p. 24.
6. *Offshore Directive - the safety of offshore oil and gas operations.* Available at http://www.hse.gov.uk/offshore/directive.htm
7. European Union, Directive 2013/30/EU of the european parliament and of the council of 12 June 2013 on safety of offshore oil and gas operations and amending Directive 2004/35/EC, *Off. J. Eur. Official Journal of the European Union* (L178/66-L178/106) (2013), p. 41.
8. *Directive 92/91/EEC - mineral-extracting industries - drilling.* Available at https://osha.europa.eu/en/legislation/directives/11
9. *European Parliament.* Available at www.europarl.europa.eu/sides/getDoc.do?pubRef=-//EP//TEXT+REPORT+A7-2013-0121+0+DOC+XML+V0//EN.
10. European Commission, *REGULATION OF THE EUROPEAN PARLIAMENT AND OF THE COUNCIL on safety of offshore oil and gas prospection, exploration and production activities*, 0309 (2011).

11. C. Frank, D.R. Eddy, G. Richard, I. Alexey and W.-O. Gerhard, *Safety Guidelines and Good Industry Practices for Oil Terminals* (2013) by United Nations Economic Commission for Europe (UNECE) Convention on the Transboundary Effects of Industrial Accidents, p. 72.
12. Storting White Paper, *An industry for the future – Norway's petroleum activities 1 Objective and summary*, 28 (2011).
13. PSA, *Regulations relating to health, safety and the environment in the petroleum activities and at certain onshore facilities (The Framework Regulations) (Last amended 17 June 2016, cf. page 4) Petroleum Safety Authority Norway Norwegian Environment Age*, (2016), pp. 1–24.
14. Directorate of Labour Inspection, *Act relating to working environment, working hours and employment protection, etc. (Working Environment Act). Arbeidslivets lover*, 64 (2012), pp. 1–64.
15. Norwegian Maritime Directorate, *Act of 16 February 2007 No. 9 relating to ship safety and security (Ship Safety and Security Act)*, 65 (2015).
16. A. Srivastava, *Ministry of Petroleum and Natural Gas Notification (Gazette of India)* (2004).
17. R. Of, I. Hydrocarbon, R. To, P. Waste, O. Pollution, T. Achieve, *Ministry of Oil - Regulations for the Conservation of Petroliam Resources Regulations for the Conservation of Petroliam Resources Ministry of Oil - Regulations for the Conservation of Petroliam Resources*, (2016).
18. A. Act, *Offshore Petroleum Act 2006* (2006).
19. C. No, *Offshore Petroleum (Royalty) Act 2006*, (2016).
20. TG.O.A. The Government of East Timor, *Timor Sea Treaty*.
21. Offshore Petroleum and Greenhouse Gas Storage Act, *Offshore Petroleum and Greenhouse Gas Storage (Safety) Regulations*, (2010).
22. *2015 Operational Review of the National Offshore Petroleum* (2015).
23. *The Oil and Gas Law Review*.
24. *Oil and gas regulation in Egypt: Overview*. Available at https://uk.practicallaw. thomsonreuters.com/7-565-7867?transitionType=Default&contextData=(sc.Default)&f irstPage=true&comp=pluk&bhcp=1
25. *Oil and gas regulation in Qatar: Overview*. Available at https://content.next. westlaw.com/Document/Id4af1a841cb511e38578f7ccc38dcbee/View/FullText. html?contextData=(sc.Default)&transitionType=Default&firstPage=true&bhcp=1
26. *Qatar National Development Strategy 2011–2016* (2016).
27. *Qatar's Legal System Governance and Business*. Available at www.nyulawglobal.org/ globalex/Qatar1.html
28. *Decree Law No. 4 of 1977 on Preserving Petroleum Resources*. Available at www. almeezan.qa/LawView.aspx?opt&LawID=2716&language=en
29. *Oil and gas regulation in the Russian Federation: overview*. Available at https:// uk.practicallaw.thomsonreuters.com/0-527-3028?transitionType=Default&contextDat a=(sc.Default)&firstPage=true&bhcp=1&comp=pluk
30. *Russia amends foreign investments regulations*. Available at www.dlapiper.com/en/ russia/insights/publications/2017/08/russia-amends-foreign-investments-regulations
31. D.K. Espinosat, *Environmental Regulation of Russia'S Offshore Oil & Gas Industry and Its Implications for the International Petroleum Market. One of the most significant events in the evolution of the petroleum-producing countries of the former Soviet Union. 2 With the*, 191 (1997).
32. *Oil & Gas Industry in India*. Available at www.ibef.org/industry/oil-gas-india.aspx
33. A.C. Anmol Soni, *The Energy and Resources Institute*, New Delhi (2014).
34. Development and Regulation, *Mines and Minerals (Development and Regulation) ACT, 1957*, Development and Regulation ACT (1957), pp. 1–36.

Global Offshore Legislations and Regulations 275

35. Development and Regulation, *The Offshore Areas Mineral (Development and Regulation) ACT, 2002*, The Offshore Areas Mineral ACT (2002), pp. 1–17.
36. *OffshoreMCR(1).pdf.*
37. S. Ladislaw, J. Nakano, A. Sieminski and A. Stanley, *U.S. Natural Gas in the Global Economy*, Retrieved April 30 from www.csis.org/features/us-natural-gas-global-economy/2018
38. *Oil Regulation.* Available at https://gettingthedealthrough.com/area/24/jurisdiction/23/oil-regulation-2017-united-states/
39. *Canada Petroleum Resources Act (R.S.C., 1985, c. 36 (2nd Supp.)).* Available at https://laws-lois.justice.gc.ca/eng/acts/C-8.5/
40. *Canada Oil and Gas Operations Act.* Available at www.neb-one.gc.ca/bts/ctrg/gnthr/cndlgsprtnct/index-eng.html
41. *Legislation and Regulations - Offshore Oil and Gas.* Available at www.nrcan.gc.ca/energy/offshore-oil-gas/5837
42. *The Atlantic Canadian Offshore.* Available at http://atlanticcanadaoffshore.ca/regulation/
43. *Health Safety and Environment Prtotecting Resources.* Available at www.saudiaramco.com.sa/content/dam/Publications/annual-review/2015/English/AR-2015-SaudiAramco-English-HSE.pdf

Index

A

Additional loads during installation and construction, 19, 22
Air compressors, 132
Air logistics, 139
All electrical control system, 178
Anchors, 35, 37, 38
Anode, 17–19
Anti agglomerates, 187
Artificial lift completion, 80
Asphaltene, 181, 190
Assessment process, 231
Australia, 257
Automated gas lift optimization in offshore, 114
Autonomous underwater vehicles (AUVs), 200
Azimuth, 59

B

Baseline, 1
Basic protection concepts, 105
Bathymetry, 3, 31
Bit, 46
Blowout prevention (BOP), 46, 69
Bottom hole assembly (BHA) & drill string, 46
Burst, 53, 54, 90

C

Canada, 270
Capacity and performance of damaged structure, 233
Casing, 44–46, 51–58, 73, 76, 80, 82, 85, 86, 89, 90, 197, 199, 228, 229
Catenaries mooring system, 38
Cathode, 17
Challenges in subsea due to water depth, 147
Chemical inhibition, 189
Chemical tank, 133
Christmas tree (X-mas tree), 89, 151
Circulating valves, 95
Classification of completions, 73
Classification of fire, 218
Classification of mooring systems, 38
Clustered well system, 157
Collapse calculations for individual casing strings, 56
Collision events, 215
Communication system, 136, 158
Completion, 45, 72, 73, 75, 76, 79–83, 86, 87, 211

Completion equipment, 89
Compliant tower and guyed tower, 27
Compliant towers, 16, 27
Concrete gravity structure, 22
Constant loads, 19
Corrosion, 17, 181, 191, 192
Corrugated plate interceptor (CPI) separators, 121
Costliest and deadliest events in oil and gas industry, 236
CPI separators, *see* Corrugated plate interceptor (CPI) separators
Crude oil heater, 133

D

Data Management and data transmission, 112
Dead weight, 35
Deep sea development options, 145
Deep steep riser, 197
Diesel system, 137
Direct access wells, 153
Direct hydraulic control system, 173
Directional drilling, 64
Drag embedment Anchor, 37
Drain header and sump caisson, 135
Drilling, 8, 9, 11, 41, 43, 44, 46, 48, 50, 57, 59, 62, 64, 65–70, 88, 124, 129, 130, 151, 153, 210
Drop-off section, 59
Dropped object, 221
Dry tree semisubmersibles, 193
Dry tree systems, 151
Dual gradient drilling, 68
Dual zone–parallel string completion, 76
Dual zone dual packer completion, 76

E

Earthquake load, 19, 21
EER, *see* Evacuation, escape & rescue (EER)
EEZ, *see* Exclusive economiczone (EEZ)
Egypt, 260
Electrical distribution manifold/module, 167
Electrical flying leads, 169
Electrical power system and communication, 171
Electrolyte, 17, 18
Emergency position indicating radio beacon, 226
Emergency position indicating radio beacon (EPIRB), 226
Emerging deepwater technologies, 193
Entertainment and recreation, 129

277

Environment dependent loads, 19, 20
EPIRB, *see* Emergency position indicating radio beacon (EPIRB)
Erosion, 90, 181, 193
Europe, 248
Evacuation, escape & rescue (EER), 222
Exclusive economiczone (EEZ), 1, 2
Expandable monobore liner extension, 197
Export pipelines, 102
Export pipelines/tankers for evacuation of oil and gas, 143
Extended life, 233
External corrosion prevention, 192

F

Factors driving deep sea development, 145
FDPSOs, *see* Floating drilling production storage and offloading systems (FDPSOs)
Field development scenarios: options/alternatives, 208
Fire, 106, 217
Fire & gas detection & safety system, 217
Fire and gas leakage protection system, 107
Fire detection systems, 218
Fire protection, 217
Fire suppression systems, 219
Fixed structures, 16, 23
Flash vessel, 121
Floating drilling production storage and offloading systems (FDPSOs), 16
Floating platform, 35
Floating platform systems, 16
Floating production, storage, and offloading systems (FPSOs), 16, 100, 124, 125, 148, 149, 153, 208, 221
Floating production systems (FPS), 16
Flow assurance, 148, 158, 179
Flow assurance challenges, 180
Fluid characterization and flow property assessments, 182
Food, 127
FPS, *see* Floating production systems (FPS)
Fresh water cooling system, 134
Fuel gas system, 129

G

Galvanic action, 18
Gas dehydration, 119
Gas detection system, 218
Gas hydrates, 180
Gas treatment, 118
Gas turbine generator, 130
Glycol reboiler, 133
Gravity platforms, 16

H

Hazards on oil and gas installations, 215
Hazardous operability (HAZOP), 107
Heating, ventilation & air conditioning equipment, 136
Helicopter incidents, 221
HDM, *see* Hydraulic distribution manifold/module (HDM)
HFL, *see* Hydraulic flying leads (HFL)
High integrity protection system (HIPS), 107
Hole cleaning & hydraulics, 46
Horizontal well completions, 80
Hot oil system, 133
HPU, *see* Hydraulic power unit (HPU)
Human factors, 213
Hybrid riser system, 195
Hydrate prevention methods, 186
Hydrate remediation, 188
Hydraulic couplers, 168
Hydraulic distribution manifold/module (HDM), 165
Hydraulic flying leads (HFL), 168
Hydraulic power unit (HPU), 134, 160, 179
Hydrocarbon releases, 215, 216

I

Ice and snow-induced load, 21
IGF unit, 121
IMO resolutions, *see* International Maritime Organization (IMO) resolutions
Impressed current cathodic protection systems, 18
Inclination (or) drift, 59
India, 266
International Maritime Organization (IMO) resolutions, 234
Infield pipelines, 101
Instrument air and utility air systems, 132
Instrumentation, remote sensing, and telemetry of real-time processes, 111
Intelligent well systems, 83
Internal corrosion inhibitors, 191
Internal corrosion prevention, 191

J

Jacketed platforms, 16, 22, 30, 33
Jack-up platform/rig, 33
Jackup rigs, 43

K

Kick-off point and build-up rate, 58
Kinetic inhibitors, 187
Kuwait, 253

Index

279

L

Lead angle, 62
LDI, *see* Low dosage inhibitors (LDI)
Life extension & assessment of offshore
    structures, 230
Life-saving equipment in EER, 223
Living accommodation, 127
Load cases, 55
Loading of tankers, 118
Logic caps, 169
Low dosage inhibitors (LDI), 187
Low pressure operations, 187

M

Main top side elements, 159
Marine growth, 21
Master/main control station (MCS), 159
Material handling, 138
MCS, see Master/main control station
    (MCS)
Measurement while drilling (MWD), 65
Medical, 128
Metallic path, 17
Meteorology, 3
Metocean, 2, 4
Mooring & anchoring, 35
Mud weight, 49
Mudmat, 163
Multilateral completions, 83
Multi-lines free standing riser, 196
Multiple quick connects, 167
Multiple zone completion, 75
Multiple zones single string completion, 79
Multiplex electro hydraulic control system, 177
MWD, *see* Measurement while drilling
    (MWD)

N

Naturally flowing completions, 79
Navigation aids, 225
Nomad systems, 201
Norway, 252
Nudging, 63

O

Offshore giant field development exercise, 209
Offshore marginal field development exercise,
    207
Offshore pipelines, 101
Offshore security threats, 240
Offshore storage, 122
Oil tankers, 123
Oil treatment, 116

P

Packers, 90
Padeyes, 163
Paraffin/wax, 181
Perforated casing completions, 75
Permanent packer, 91
Pile, 38
Piloted hydraulic control system, 174
Pipelines, 124
Platform wells/dry trees, 100
Potable water system, 133
Power generation, 129
Pressure testing conditions, 56
Procedural aspects related to safety, 216
Process leaks, 216
Process safety & hydrocarbon releases, 216
Processing in offshore, 116
Processing platforms, 101
Produced water treatment, 120
Production, 4–6, 52–54, 57, 67, 75, 99, 100, 103,
    106, 116, 145, 147–149, 157, 171, 193,
    208, 248, 264, 266
Production tubing, 89
Production well templates, 157
Proof of structural integrity with increased
    loads, 232
Protection of offshore facilities/rigs, 235
Proximity (anti-collision) analysis, 64
Pyrotechnics, 227

Q

Qatar, 261

R

Recoverable reserves, 145
Reservoir production characteristics, 148
Reservoir structure, 148
Retrievable packer, 90
Rig equipment, 46
Riser leaks, 216
Risks, 145
Russia, 263

S

Sacrificial anodes, 18
Safety in logistics operations related to offshore
    installation, 220
Safety valves, 93
Sand production, 51
SART, *see* Search and rescue transponder
    (SART)
Satellite well system, 155
Saudi Arabia, 272

280 Index

SCADA, *see* Supervisory control and data acquisition system (SCADA)
Scale management, 192
Scales, 181
SDS components, *see* Subsea distribution system (SDS) components
Sea logistics, 139
Sea water injection, 121
Seabed separation, 203
Search and rescue transponder (SART), 226
Selection of hydrate control method, 189
Sequenced hydraulic control system, 175
Severe slugging, 182
Sewage treatment system, 138
Shallow *vs.* deep flow assurance scenario, 180
Ship-shaped floating production, storage, & offloading systems, 16
Shut down panel, 105
Shut in, 107
Single buoy mooring, 125
Single point mooring system, 39
Single zone completion, 75
Skimmer vessel, 133
Slugging, 182
Smart well technology, 199
Smoking and alcohol, 129
Spar platforms, 16
Stand-alone development, 155
Steady-state hydraulic and thermal performance analyses, 183
Structural collapse, 230
Subsea accumulator module, 169
Subsea completion, 86
Subsea control module (SCM), 169
Subsea distribution assembly, 165
Subsea distribution system (SDS) components, 163
Subsea elements, 162
Subsea field development, 149
Subsea monitoring, control & communication system, 158
Subsea multiphase pumps, 202
Subsea power supply, 179
Subsea pressure boosting, 204
Subsea processing, 202
Subsea production control system, 171
Subsea tie-back development, 153
Subsea umbilical termination assembly (SUTA), 163
Subsea well completion, 151
Subsea wells clusters, 153
Subsea wells/wet trees, 100
Suction anchor, 38
Sump caisson, 135
Supervisory control and data acquisition system (SCADA), 95, 100, 109, 111, 112, 114, 116, 212

Surface facility protection, 107
SUTA, *see* Subsea umbilical termination assembly (SUTA)
System design and operability, 184
System safety, 216
System shut down, 184
System start up, 184

T

Tangent section, 59
Technology development, 107
Temperature induced load, 19, 21
Template well system, 156
Tension leg platforms (TLP), 16, 193
Thermal & chemical wax dissolution, 190
Thermal control, 189
Thermal insulation and heating, 187
Thermodynamic inhibitors, 186
TLP, *see* Tension leg platforms (TLP)
Topside umbilical termination assembly (TUTA), 160
Total horizontal deviation, 62
Trainings, 244
Trajectory, 45, 57, 59, 61, 62
Transducer/sensor, 171
Transient flow hydraulic and thermal performance analyses, 185
Transmission/cross country pipelines, 101
Transportation of oil and gas, 123
Tubing connections, 90
Tubing grade, 89
Tubing size, 89
Tubing weight, 90
TUTA, *see* Topside umbilical termination assembly (TUTA)
Types of control systems, 173
Types of directional patterns, 61
Types of directional wells, 64
Types of dual gradient drilling, 70
Types of subsea completions, 87
Typical flow assurance processes, 182

U

Umbilical, 163
Umbilical termination head (UTH), 165
UNCLOS, *see* United Nations Convention on the Law of the Sea (UNCLOS)
Unforeseen loads, 19, 22
United Nations Convention on the Law of the Sea (UNCLOS), 247
USA, 269
UTH, *see* Umbilical termination head (UTH)
Utility water system, 135
Utility/diesel generators, 130

Index

V

Variable loads, 19
Vertical load anchor, 38
Vessel collisions, 221

W

Water depth, 147
Water removal, 187
Wave, 3, 19, 20
Wax control guidelines, 189

Wax management strategy, 189
Well automation, 114
Well control& protection, 109
Well groupings, 155
Well integrity, 228
Wellbore, 46–48
Wellhead, 46, 112, 150
Wells (subsea/platform wells), 100
Wet tree systems, 100
Wind, 20, 21
Workflow automation, 11

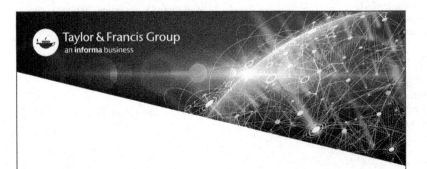